BREAKING BOUNDARIES

The Science of Our Planet

BREAKING BOUNDARIES

The Science of Our Planet

OWEN GAFFNEY AND
JOHAN ROCKSTRÖM

Editor Becky Gee	**Jacket Designer** Akiko Kato
Art Editor Mark Lloyd	**Art Director** Karen Self
Managing Editor Angeles Gavira	**Associate Publishing Director**
Managing Art Editor Michael Duffy	Liz Wheeler
Production Editor Kavita Varma	**Publishing Director**
Senior Production Controller	Jonathan Metcalf
Meskerem Berhane	

Colour visualizations by Félix Pharand-Deschênes, Globaïa

Documentary film produced by Silverback Films Ltd

First published in Great Britain in 2021 by
Dorling Kindersley Limited
DK, One Embassy Gardens, 8 Viaduct Gardens,
London, SW11 7BW

The authorised representative in the EEA is
Dorling Kindersley Verlag GmbH. Arnulfstr. 124,
80636 Munich, Germany

A CIP catalogue record for this book
is available from the British Library.
ISBN: 978-0-2414-6675-9

Printed and bound in Latvia

For the curious
www.dk.com

This book was made with Forest Stewardship
Council ™ certified paper – one small step in
DK's commitment to a sustainable future.
For more information go to
www.dk.com/our-green-pledge

ABOUT THE AUTHORS

OWEN GAFFNEY

Owen Gaffney was born at 324 parts per million carbon dioxide. He is a global sustainability writer, analyst, and strategist, based at the Potsdam Institute for Climate Impact Research and Stockholm Resilience Centre. He is a Future Earth Senior Fellow and an Edmund Hillary Fellow (New Zealand), and also sits on the faculty of Singularity University. He cofounded the Exponential Roadmap Initiative and the film company Gaiaxia and advises organizations such as the Global Commons Alliance. Owen trained in astronautic engineering, film-making, and journalism. His work has appeared in academic journals such as *Nature* and *Science,* and he has contributed to *New Scientist* and the WWF's Living Planet Report. He cofounded the Future Earth Media Lab and *Rethink* magazine and sits on the editorial board of the *Anthropocene* magazine. Owen lives in a forest on the Stockholm archipelago.

JOHAN ROCKSTRÖM

Johan Rockström was born at 320 parts per million. He is a professor in Earth system science at the University of Potsdam and professor in water systems and global sustainability at Stockholm University, as well as the director of the Potsdam Institute for Climate Impact Research, founder of the Stockholm Resilience Centre, and chief scientist at Conservation International. He is an Earth system researcher, whose work has appeared in high-impact academic journals such as *Nature* and *Science*. He has given four TED talks on the state of the planet. Johan is co-chair of the Earth Commission and Future Earth, an international research network. He is regularly invited to speak at the World Economic Forum and has contributed to WWF's Living Planet Report. Johan lives in Berlin.

CONTENTS

ACT III

FOREWORD

GRETA THUNBERG
SWEDISH ENVIRONMENTAL ACTIVIST

A stable planet is a necessary condition for the well-being of our civilization. And a stable planet for life as we know it requires an atmosphere that does not contain too much greenhouse gas. This is basic science.

The safety limit for the level of carbon dioxide in the atmosphere is thought to be around 350 parts per million. We reached that landmark sometime in 1987, and in 2020 we surpassed 415 parts per million. The world has not experienced such high levels of atmospheric carbon dioxide for at least 3 million years. This increase is happening at an unprecedented speed. Half of the anthropogenic carbon dioxide emissions have been emitted since 1990 – in the past 30 years. As a result, our functioning and balanced atmosphere has become a finite natural resource. A limited resource that today is being used up primarily by a very small number of people. This is the heart of the problem: climate injustice. It is not only an issue between nations; it is an issue within all societies.

The richest 10 per cent of the world's population emit more carbon dioxide than the remaining 90 per cent. On average, the top 1 per cent of income earners emit 74 tonnes (81 tons) of carbon dioxide per person every year. For the 50 per cent of the world population with the lowest incomes, that same per capita figure is 0.69 tonnes (0.76 tons). These high emitters are the people we consider to be successful. They are our leaders, our celebrities, our role models. The people we aspire to be like. Or just about anyone with a high enough income. There are many elephants in the room when it comes to the climate crisis, and climate injustice is undoubtedly one of the biggest.

Many people say that we should not turn the climate crisis into a moral issue. This will only result in inflicting guilt and shame, and that will be counterproductive. And yet, the Paris Agreement of

2015 – the only functioning tool we have at our disposal today – is a treaty completely founded on equity and morality, with all its non-binding and voluntary goals and targets.

Furthermore, the school strike movement is based firmly on the idea of climate justice. On morals. Or guilt, if you so wish. And the reason the movement has been so successful is probably because the dire consequences of climate inaction no longer just threaten people "far away" – now they involve everyone's children. The consequences have moved closer to those of us who had until this point felt fairly safe. People have started to get scared. Scared as in really frightened for the well-being and safety of their loved ones.

If lowering our emissions of greenhouse gases were one of humanity's main objectives, we could start right away. When we humans really put our minds and resources into something, we can achieve almost anything. Just take the quick and effective development of vaccines against COVID-19. Unfortunately, though, our objective today is not to lower our emissions. That is not what we are fighting for. We are fighting to maintain our way of living.

"The American way of life is not up for negotiations. Period," said US President George H. W. Bush just before the 1992 UN Earth summit in Rio de Janeiro. Since then, not much has changed. Most world leaders would probably still say the same, if not through their words, then certainly through their actions. Or inactions, rather.

So, instead of lowering our emissions we search for "solutions". Solutions to what, we should ask? Solutions to a crisis that the vast majority of us cannot even begin to fully understand? Or solutions that allow us to be able to go on like before?

Well, why not both you say? But the sad answer is that we have left it too late for that to be possible. If you read the current best available science reports – the Intergovernmental Panel on Climate Change Special Report on Global Warming of 1.5°C (SR1.5), the Emissions Gap Report produced by the United Nations Environment Programme, and the Global Assessment Report on Biodiversity and Ecosystem Services produced by the Intergovernmental Science-Policy Platform on Biodiversity and Ecosystem Services – you will see that the climate and ecological crisis can no longer be solved within today's financial or legal systems.

For us to stay below 1.5°C (2.7°F) global warming, or even well below 2°C (3.6°F), we will need to start tearing up valid contracts and deals. We will need to leave assets in the ground. We will need to boost every imaginable carbon sink to the very maximum. A standing tree must then be worth more than a dead one. We will need to transfer to a zero carbon society while leaving no one behind in the process. And that is no longer possible within today's societies. The need for system change is no longer an opinion; it is a fact.

The climate and ecological crisis cannot be solved by individual changes alone. Nor by the "market" for that matter. We need huge, broad, political and structural changes on an unprecedented scale, "in all aspects of society" to quote the Intergovernmental Panel on Climate Change. However, those changes are nowhere in sight today. Nor are they likely to appear anytime soon.

When the first vaccines for COVID-19 were rolled out in December 2020, health services across the globe called upon celebrities to receive some of the first shots, as a way of convincing others to do the same. This is a well-proven method; it is how we humans work. We are social animals – herd animals if you like – we follow our leaders and copy the behaviour of those around us.

According to the latest Emissions Gap Report, "Popular debate has often pitted 'behaviour change' and 'system change' against each other, presented as a trade-off between two choices. . . . however, system change and behaviour change are two sides of the same coin."

The climate and ecological crisis is, in fact, only a symptom of a much larger sustainability crisis. A crisis for which there will be no vaccine. It is a crisis that involves everything from climate and ecological breakdown, loss of fertile soils and biodiversity, and acidification of the oceans, to loss of forests and wildlife and – indeed – emergence of new diseases and new pandemics.

During the five years since the Paris Agreement was signed, a lot has happened. But the action needed is still nowhere in sight. The gap between what we need to do and what is actually being done is widening by the minute. We are still speeding in the wrong direction.

Commitments are being made, distant hypothetical targets are being set, and big speeches are being given. Yet, when it comes to the immediate action we need to take, we are still in a state of complete denial, as we continue to waste our time creating new loopholes justified by empty words and creative accounting.

The prospect of solving the climate and ecological crisis is not very hopeful, to say the least. This bleak outlook is probably part of the reason why the situation is so incredibly hard to communicate in our age of social media "likes" and celebrity culture. Nevertheless, we have to tell it like it is, because only then can there be true hope. We should be adult enough to handle the truth. And until we are, the children will continue to be the adults.

The science dictates that we now need to do the seemingly impossible. And that is sadly no longer a metaphor. That is where we are. So, instead of focusing on vague, incomplete, insufficient hypothetical targets set up by people in power, we should put all our efforts into communicating the reality. Because, despite what you may think, the vast majority of us are not aware of the situation we face. Or rather, we have not yet been made aware.

And that's where the true hope lies. How, you may ask?

Well, just imagine how many leading politicians, newspapers, TV stations, public figures, celebrities, and influencers there are in the world today. And imagine if just a small number of them started to treat the climate crisis like the crisis it is. Imagine who they might reach. Imagine if millions and millions of everyday people like you and me were really aware of what we are up against. Then everything would change overnight. We can still do this – but only if we speak the truth.

People often ask me if there is one thing that they can do that will really make a difference, and if so – what is it? My answer is always: inform and educate yourself as much as you possibly can and then spread that awareness to others. Because once you understand the full meaning and real consequences of our situation, then you will know what to do.

Our hope lies in the facts and the knowledge that make up the current best available science, and that this knowledge can spread fast and far enough. That's where you come in. And that's what reading this book is all about.

January 2021

INTRODUCTION

Dear friends,
Humanity is waging war on nature.
This is suicidal.

ANTÓNIO GUTERRES
UNITED NATIONS SECRETARY-GENERAL

It is night. You are driving hard down a steep, winding road. No barriers or guard rails protect you or provide any warning you are careening too close to the edge. The headlights flicker out. At any moment, the car could skid off the road and fly into the ravine, where the vehicle and its occupants would rapidly change state. The children in the back are screaming.

You might think the narrow road, the murky darkness, and steep cliffs would force you to slow down, but instead you bump along in the dark, taking hairpin bends at speed.

This may seem like a nightmare, but humanity is taking the same risk with our planet and our common future. Earth's life support system is the sum of the planet's ice sheets, oceans, forests, rivers and lakes, and rich diversity of life, as well as gigantic recirculations of carbon, water, nitrogen, and phosphorus. This system is now most definitely unstable. At any moment, we could push it over the edge, taking us – 7.8 billion people – with it.

In the past few decades, scientists have been frantically trying to work out how far our life support system can be pushed before everything we know and love starts to break down. Ten years ago, this research allowed us to estimate, for the first time, where to build the safety barriers to protect us from falling off the cliff. We called these the "planetary boundaries". They define, scientifically, a safe operating space on Earth for us humans to have a good chance of a thriving future. After all, every child's birth right is a stable, resilient planet. Since the dawn of civilization, 10,000 years ago, this has been our common heritage. If Earth remains within the planetary boundaries, we have a better chance of a long stable future. Outside the boundaries, anything could happen. We have written this book to tell this story.

Time is running out. The decade we have just stepped into – the roaring 2020s – will be decisive for humanity. It is the moment to catalyse the most remarkable transition in history to become effective stewards of Earth. The scale of the challenge is immense. In the same way that the 1960s had the Moonshot, the 2020s has the Earthshot. The goal of the Earthshot is nothing less than stabilizing our planet's life support system. But compared with landing men on the Moon, the stakes are far higher.

If we can achieve this goal, and it is a big "if", then perhaps it will mark a profound transition for Earth, in which one species gains the power to deliberately and positively influence a planet's habitability. We are far from that point right now. In fact, at the moment, we are acting as if our aim is to be one of the select few species that have alone managed to unwittingly disrupt the habitability of a planet. This is occurring despite our vast knowledge, our global political systems, and our advancing technology. We have already struggled to contain a global pandemic, struggled to slow extinction rates, struggled to halt deforestation, and struggled to rein in greenhouse gas emissions. We have cut a gaping hole in the ozone layer, and we watch mutely as corals bleach, ice sheets melt, and fires blaze. Our power is outstripping our wisdom.

But things are changing. In the past decade, we have seen an explosion of evidence that a global population of 7.8 billion people, growing to 10 billion, can live a good life within Earth's boundaries. And this can be achieved by 2050. The big drama is that, in order to succeed, we need to do much of the work by 2030 or thereabouts, which means we need to start now.

The Earthshot comes at a time of deep dysfunction in society. Imagine if NASA found an asteroid that was on a collision course with Earth. Estimated time of arrival: one decade. What would we do? Would we use our superpower – cooperation – to come together and combine our economic resources and our best minds to solve the problem? Or would we do nothing? Would we hope that we get lucky and the rock grazes past?

Actually, we do not need to imagine. A pandemic swept across the world in 2020, in the midst of writing this book. COVID-19 was no black swan – a rare and large-scale, unforeseeable event. For at least a decade, scientists had been sending unequivocal warning signals specifically about a coronavirus. Health experts reckoned the cost of an adequate early warning system amounted to one or two US dollars per person on Earth every year. But governments failed to heed the warnings and within months of the outbreak more than half the population on Earth was in lockdown. We are ignoring knowledge of existential risks.

Pandemics, climate change, and mass extinctions of life on Earth are part of the same pattern and they are tightly connected. We are now living in a new geological age: the Anthropocene. This new age is characterized by speed, scale, surprise, and connectivity. Risks are amplified because we separate our economy from nature and humanity. The pandemic recovery is a transformative moment to rethink our relationships with the economy, with each other, and with our planet.

Time is precious, so in this book we want to concentrate on solutions. We have identified six system transformations – energy, food and land, inequality and poverty, cities, population and health, and technology – that are necessary to stabilize Earth and ensure economic security and prosperity. Perhaps surprisingly, we are optimistic that these system changes are within reach. This is, in part, because four positive forces are aligning:

1. Social movements. The speed, scale, and impact of campaigns such as FridaysForFuture are astounding. They have changed the conversation and upped the urgency. These movements are calling out and exposing political and industrial failures. Their impact cannot be underestimated. And they are not alone. Business leaders, investors, and legal experts are throwing their weight behind the mission.

2. Political momentum. The global economy is driven by three economic powerhouses: Europe, China, and the United

States. In 2019, Europe committed to reach net zero greenhouse gas emissions by 2050. In September 2020, Chinese president Xi Jinping announced his country would be carbon neutral by 2060 at the latest. Then Joe Biden came to power and committed to putting the United States on course to reach net zero by 2050 (and, incredibly, to run on 100 per cent clean electricity by 2035). Like the G7 and G20 groups of countries, this creates a new "G3 for climate", with the three largest economic regions driving a remarkable and necessary economic transformation, which will impact the whole world.

3. Economic momentum. The fossil fuel era is over. Solar power is now the cheapest source of electricity in the history of humanity.

4. Technological innovation. The fourth industrial revolution, from 5G to artificial intelligence and biotechnology, is on the cusp of disrupting all economic sectors. This can and must be targeted at supporting the economic transformation.

Combined, these forces are pushing us inexorably towards a positive tipping point. Planetary awareness is finally emerging. Of course, we acknowledge that this is just the start. Political leaders must commit to even more ambitious goals.

The pandemic has provided a unique opportunity to fix our broken economic system and re-evaluate what really matters. The fall in air pollution during the lockdown months gave a glimpse of a future without dense smog hanging over cities. The crisis often brought out the best of humanity. Let's hang on to that. But let's also ask, what do we really value? What kind of societies do we want to live in? Can we build an economy that does not fall to pieces at the slightest sign of uncertainty? And can we build an economic recovery over the next decade that simultaneously supports a resilient planet? We need to jettison old school economic thinking, go back to the classroom, and learn a new set of three "R"s: resilience, regeneration, and recirculation. Economic growth at the expense of Earth's biosphere – our living planet – must be redirected towards growth based on knowledge, information, digitalization, services, and sharing. This is the Anthropocene school of economic thinking.

What if we fail to make enough progress in the next decade? Do we fall off the cliff? Even if greenhouse gas emissions are reduced by 50 per cent this decade, the risk of skidding over the cliff remains high. But we would not fall over the edge in 2031; the danger is that we trigger self-warming and make the drift unstoppable. Here are some of the risks we face: irreversibly losing tropical coral reefs; pushing the Amazon rainforest across a tipping point; setting Greenland and West Antarctic ice sheets on an irreversible melting course; and sparking unstoppable methane release from thawing permafrost. There will never be a point where all is lost. Many of our colleagues rightly say that if we cross tipping points we may still be able to control the rate of change by acting decisively, but the longer we wait, the more turbulence and turmoil we leave to our children and grandchildren for generations to come. And if there is still a chance to pull us back from the brink, should we not put all our efforts into that action?

This book tells the story of humanity's journey towards planetary stewardship in three acts. Ultimately, it is about changing course, about responsibility and opportunity, about setting a new standard for our accomplishments.

Act I is the story of Earth's life support system: the cascading water, carbon, nitrogen, and phosphorus cycles; the crashing continents; the rolling waves and receding ice sheets; and the twists, turns, and surprises of evolution. Here, we will explain that Earth has a hair trigger: a small change could set off a cascade of tipping points that could send global temperatures soaring. Act I will also tell the story of how one species, through its own revolutions – agricultural, scientific, and industrial – reshaped our planet. Now, this one species is hammering Earth's hair trigger from all sides.

Act II is the story of the remarkable scientific achievements of the past three decades. Scientists have been racing to assemble an unparalleled understanding of the health of the planet. This has led to the undisputed conclusion that the rate of change of Earth's life support system is accelerating. The only stable state that we know can support civilization – the Holocene – is disappearing in the rear-view mirror. This is a planetary emergency.

Act III is the story of the Earthshot – our most important mission. Humanity simply must become good stewards of Earth or we will not be around for very long. Our decisions in the next ten years will influence the next ten millennia.

When assembled, the fragmented knowledge about our planet and our place in it amounts to nothing less than a paradigm shift in our understanding of Earth and our responsibility in relation to it. How will this translate to a cultural, economic, and political paradigm shift? The way we operate as societies must fundamentally transform – this is revolution not evolution. Belief in infinite growth based on material extraction and expulsion is not compatible with a long-term civilization here on Earth. But there is another way. Our most important insight in this book is its simplest: it is past time we became planetary stewards. *Breaking Boundaries: The Science of Our Planet* is about that transformation. It is not about where we are coming from. It is about where we are going.

It took 50 years for us – and let's be clear, primarily the "us" here refers to the wealthiest people in the wealthiest countries – to push Earth beyond a 10,000-year period of extraordinary stability. The decisions we make today – literally *today* – this decade and in the next 50 years will influence the stability of our planet for the coming 10,000 years.

Three things unite us: our common human identity within a truly global civilization, our common home – this planet we call Earth – and our common future. If we can turn our ship within a decade, if we become effective planetary stewards, then maybe, just maybe, we will have earned our name *Homo sapiens* – wise man.

This leaves us with a final question. What kind of world do we want to leave to our children? Let's leave our children nothing. No greenhouse gas emissions. No biodiversity loss. No poverty. This is not a manifesto, nor is it an aspiration. It is one of two choices that we – humanity – have on the table right now.

January 2021

ACT

I

THREE REVOLUTIONS THAT SHAPED OUR PLANET

Look again at that dot. That's here. That's home. That's us. On it, everyone you love, everyone you know, everyone you ever heard of, every human being who ever was, lived out their lives.

CARL SAGAN
*PALE BLUE DOT: A VISION
OF THE HUMAN FUTURE IN SPACE,* 1994

In 1990, NASA snapped a photograph, a kind of family portrait, from the Voyager I spacecraft, which had just reached the edge of our solar system. For the final time on the mission, the camera pointed back to Earth, 6 billion kilometres (3.7 billion miles) away. This photograph became known as the "Pale Blue Dot". Since the image was taken, 30 years ago, scientists have identified more than 4,000 exoplanets (planets orbiting other stars). Some are planets like ours.

Each year, our understanding of these exotic exoplanets grows. We can figure out if they orbit within the habitable zone of a star; measure their density; calculate how much they weigh. That sort of thing. In the next decade, we may even be able to elucidate the composition of their atmospheres and whether they might have liquid water on their surfaces. Isn't it remarkable that without leaving Earth, we will soon be able to get a glimpse of the potential life support systems of distant planets orbiting strange stars?

Since the "Pale Blue Dot", any alien life form able to observe Earth from a distant exoplanet might notice profound changes afoot. A leap in the amount of greenhouse gases in the atmosphere. Dead zones spreading out over the ocean. Ice sheets collapsing. Sea levels rising. A colossal shift in the acidity of the ocean. For any observer accumulating large amounts of data over the past century, it is impossible to miss the fact that Earth's life support system is undergoing some sort of trauma. They would probably conclude that, yes, life exists on Earth – indeed, there is a rich biosphere – but the planet is likely in the midst of some kind of mass extinction. They might wonder why, and what will happen next. Some of the inhabitants of Earth are thinking the same thing.

To answer this question – what happens next? – we must first understand the risks to the very stability of Earth's life support systems. With these insights, we can explore what we can do about them and seek appropriate pathways towards a prosperous and equitable future on our beloved little Pale Blue Dot.

The universe is 13.8 billion years old and expanding like a balloon. Within it hurtle 2 trillion galaxies (at least). At the centre of one, the Milky Way, a supermassive black hole as large as 4 million Suns spins. This black hole, called Sagittarius A, holds between 100 and 400 billion stars in its grip. On an outer arm, one star hurtles through space dragging with it eight planets. Just one of these planets, Earth, is known to have life.[1] It orbits near the edge of the habitable zone closest to the Sun.

What do we know about life on Earth? We have a pretty good grip on some of the vital statistics. The most numerous organism on Earth is probably the bacterium *Pelagibacter ubique*, found in the ocean and fresh water. This was first described in 2002. Scientists estimate that the ocean contains 10^{29} microbes, far more than the 10^{22} stars in the observable universe. If you were to place all life

---

[1] Other parts of the solar system are potentially habitable. Prime suspects are Saturn's moon Enceladus and Jupiter's Europa. Beneath their ice shells, oceans may lie.

on Earth on a weighing scale, the writhing, wriggling mess would weigh 550 billion tonnes (600 billion tons). Plants make up 82 per cent of this weight. There are about 3 trillion trees on Earth, but we are losing 15 billion a year through deforestation. We have cut down 46 per cent since the dawn of agriculture 10,000 years ago, and most of the chopping happened in the past two centuries. We have been busy.

If we remove from the weighing scale all life that is not mammalian – all species of birds, reptiles, amphibians, plants, molluscs, insects, and arachnids – a full 96 per cent of the weight of mammals will be made up of humans, and cows, sheep, pigs, and horses. Wild mammals – big cats, rodents, blue whales, dolphins, and the other 6,500 species of generally furry creatures – make up just 4 per cent. It was the other way around a few short centuries ago.

Earth 1.0 to 4.0

Some 4.5 billion years ago, the planets in our solar system began forming from the hot mess of gases and rocks spinning around the Sun. Earth 1.0's defining features included constant bombardment by meteorites, no ocean, no life. Three revolutions followed: simple single-celled life, Earth 2.0; the evolutionary leap to photosynthesis with oxygen, Earth 3.0; and complex life, Earth 4.0.

These revolutions have several things in common. They are all linked to evolutionary innovation and they all opened up new ways of using energy. Each step led to a rise in complexity and radically altered the amount of information processing occurring on Earth. This meant that organisms new to the scene could use and spit out resources faster. After each revolution, things eventually settled to a new order, but not without consequences. Each revolution disrupted some or all of Earth's natural cycles of carbon, oxygen, water, nitrogen, or phosphorus, often for millions of years. Eventually, some sort of harmony was restored when new organisms evolved to recycle the waste materials. The relative stability of Earth's life support system – the rivers, lakes, oceans, soils, rocks, and air, as well as vast cycles of water, carbon, oxygen, nitrogen, and phosphorus – depends on the emergence of circular flows and constant recycling.

Earth is now undergoing another evolutionary leap and the rate of change of Earth's life support system is accelerating. This time, one species is driving it: us.

So, how do we know any of this? The sum total of our knowledge about everything, from expanding universes and folding proteins to the origin of humanity and the data showing that women smile more than men,[2] is contained in the 80 million research papers that have been published since the first scientific journals were established in Paris and London in 1665.[3]

Scientific knowledge is expanding like the universe. In the region of 3 million academic papers are published every year (approximately one every 10 seconds). More than 70,000 papers have been published on a single protein, called "p53", and there are more than 10,000 papers on algorithms for self-driving cars. With so much information flying around in so many disparate journals, largely locked behind paywalls, research can seem confusing, fragmented, unstructured, and impossible to keep up with, even for people working in science. While our collective knowledge has never been greater, at an individual level scientists are specialists and we organize information for specialists. Throughout this book, we will pull out the essential scientific findings to create a big picture narrative as we move through Earth's history, from a lifeless lump of rock to a momentous event around 3 million years ago: the onset of the ice ages that shaped us as a species.

Let's start with age. How on Earth do we know that Earth is 4.5 billion years old?

Earth 1.0: Unravelling our earliest origins
In the 4th century BCE, Aristotle mused that Earth had existed eternally, probably thinking to himself that this would be difficult to disprove. Charles Lyell, an influential 19th-century geologist, tended to agree, maybe for similar reasons.

---

2 It is probably more to do with culture than genetics, according to a study of 162 research papers on smiling.
3 The first academic journal, *Journal des Sçavans*, was published in Paris in January 1665, followed by *Philosophical Transactions of the Royal Society* in London in March 1665.

One of the first to take a heroic stab at the problem of Earth's age was the Roman orator Julius Africanus, during the reign of Emperor Nero in the 3rd century CE. Africanus composed a masterwork: a monumental five-book world history. He scoured scrolls and tablets written in Hebrew, Greek, Egyptian, and Persian to exhaustively chart the chronology of Earth. With great fanfare, Africanus announced Earth to be 5,720 years old. There were very few quibbles with this estimate for a good 15 centuries. Even as late as 1650, Ireland's Archbishop James Ussher proposed Earth came into existence on 22 October 4004BCE, a rather alarmingly precise date not so far removed from that of Africanus. Indeed, both estimates give a pretty good start date for the age of recorded history and the formation of civilizations, give or take a thousand years.

By the late 19th century, some geologists were forming a different conclusion: perhaps Earth could be around 100 million years old. But then the discovery of radioactive material in rocks changed all that. In 1956, Clair Cameron Patterson, a geophysicist at the California Institute of Technology, published one of the most remarkable papers in the scientific canon, "Age of meteorites and the Earth".[4] Patterson realized that the mineral zircon, found in the oldest igneous rocks throughout Earth's crust, could unlock the age of the planet. Zircon is never pure when it forms; it is shot through with tiny uranium particles. Patterson knew that uranium has a reliable habit of decaying to the metal lead. If you find any lead present in a zircon sample, it is because of radioactive decay. If you can measure the amount of lead carefully, you can work out exactly when the zircon formed and hence the age of the sample. Patterson used this information to peg Earth at 4.55 billion years old, give or take 70 million years. Since then, there have been minor quibbles about the precise age of our planet, but no one has proposed any strong evidence for a radically different figure. This marks the beginning of Earth 1.0 – the Hadean – and the formation of the planet. This first state ended with a more ordered, cooler planet with an ocean and atmosphere.

---

4 Academics often seem locked in an infinite arms race to think up underwhelming titles for research papers. Albert Einstein's 1905 masterpiece showing $E=mc^2$ was dryly titled "On the electrodynamics of moving bodies".

Earth 2.0: Simple single-celled life

The first planetary revolution was the evolution of life on Earth 3.8 billion years ago, and perhaps earlier, although this is where things get contentious. Essentially, as far as we can tell, single-celled life known as archaea evolved soon after the ocean formed on Earth. At this moment, Earth 1.0, the lifeless hellish Hadean, ended and the planet entered a new geological aeon:[5] the Archaean. Earth 2.0 had arrived.

Unlike all life around us, archaea do not need oxygen to survive. In fact, oxygen can kill them. This was not a problem in the Archaean aeon because the planet was devoid of oxygen in the atmosphere. About 2.5 billion years ago, that changed. The second revolution in our planet's evolution occurred when a new life form evolved in the ocean: cyanobacteria, sometimes called blue-green algae. For the first time on Earth, these microorganisms took energy from the Sun and emitted a waste product: oxygen. Photosynthesis that produced oxygen had arrived.[6] Goodbye Earth 2.0. Hello Earth 3.0, the Proterozoic aeon. This name originates from two Greek words: *protero*, meaning "early", and *zoic*, meaning "life".

Earth 3.0: Photosynthesis with oxygen

Cyanobacteria kick-started one of the most profound and catastrophic events in our planet's history: the Great Oxygenation Event.[7] This tiny organism's waste changed the composition of the early atmosphere. Oxygen levels began to rise. An ozone layer formed, which stopped harmful radiation from the Sun reaching Earth's surface, thus paving the way for life on land (but don't get too excited; we are still 2 billion years away from that).

The old order fell away. Archaea's dominance declined as they retreated into deep cracks and fissures in rocks without oxygen and below the ocean. Some, though, evolved to thrive in this new

---

5 Aeons are the largest divisions of geological time. There are just four, which are subdivided into eras, periods, and epochs.
6 It is likely that a form of photosynthesis existed before this time, but, critically, it did not release oxygen.
7 This evolutionary leap wiped out more than 75 per cent of life, making it a "mass extinction event" in biodiversity jargon.

environment. Cyanobacteria are one of only two species on Earth that are known to have fundamentally altered the functioning of the Earth system on their own. We will get to the second – *Homo sapiens* – in approximately 2.5 billion years.

The arrival of cyanobacteria coincided with a colossal shift in the climate, and Earth plunged into a deep freeze, deeper than anything imaginable. Earth became a snowball.

At this point, we should note that Earth's thermostat has three main settings:

- The hothouse, where there is no ice at the poles and little or no ice anywhere on Earth.
- The icehouse, where it is cool enough to have ice at the poles, like today, and indeed the past few million years. The icehouse has two stable states: deep, cold, long ice ages and shorter warm periods called interglacials.
- The snowball, where ice creeps further towards the equator, locking the entire planet within an ice shell at least 1 kilometre (0.6 miles) thick in places. During snowballs, life still clings on around remote hydrothermal vents on the ocean floor.

Earth's three climate states

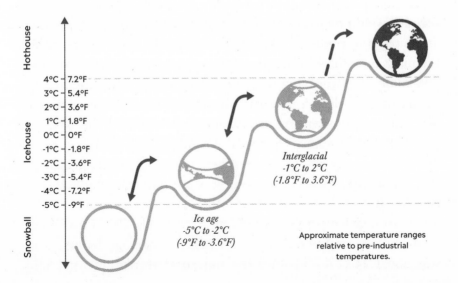

Think of these three states as an oven on low heat, a refrigerator with the door either closed or left open, and a deep freezer. The hothouse has dominated for much of Earth's history, but at the start and end of Earth 3.0, the planet snowballed.

Some scientists think the hard, impenetrable snowball state is unlikely and it might have been more like a slushball, with a little open water at the equator. Either way, this was a trying time for life. Some academics dispute whether snowball or slushball are even feasible, because it seems unlikely that a planet could escape such an icy death spiral. These critics have a point. The white surface of ice reflects heat back to space, thus cooling the planet. As ice creeps towards the equator, it covers a greater surface area of Earth, reflecting more heat away. This leads to more cooling, which creates more ice: a vicious cycle. Could we reach a point where defrosting becomes impossible? If so, we are faced with runaway climate change taking hold and locking the planet into a bleak ice prison.

In 1992, Joe Kirschvink at the California Institute of Technology identified an escape route. He noted that volcanic activity, such as hydrothermal vents or undersea volcanoes, would still slowly spew a little carbon dioxide even during snowballs. This greenhouse gas would gradually seep into the atmosphere. Although the levels would be relatively low, over millions of years carbon dioxide would build up. Eventually, enough would accumulate to trap what little heat was attempting to escape the planet's icy clutches. Ice at the equator would melt, thus setting off a chain reaction. The dark surface beneath would be exposed and would absorb more heat, creating an amplifying feedback loop or virtuous cycle, warming the area around it until the ice retreats and we are back to the icehouse, and potentially on a trajectory taking Earth to a hothouse state.

Carbon dioxide is a powerful absorber of heat. We have known this since 1859, when the prominent Irish scientist John Tyndall discovered that some "perfectly colourless and invisible gases and vapours" absorb heat while others do not. He later speculated that the ability of carbon dioxide to store heat may be linked to climate. Indeed, without it and other greenhouse gases Earth's global average temperature would drop to a chilly -18°C (-0.4°F), the temperature of a freezer. Instead, the global average temperature was about 14°C (57°F) until the industrial revolution went into overdrive in the 1950s. Throughout this book, we will talk about global warming relative to the industrial revolution baseline.

The last snowball ended 635 million years ago. Combined with high levels of oxygen in the atmosphere, the extreme shift out of the snowball appears to have created the perfect conditions for genetic diversity. Fossils show that the diversity of life on Earth exploded some 542 million years ago. Known as the Cambrian explosion, this marked the end of Earth 3.0 and kick-started Earth 4.0: complex life.

Earth 4.0: Life explodes

For 100 million years after the Cambrian explosion, life steadily increased in complexity and diversity, with rapid diversification in body shape and size, organs, skeletons, backbones, circulation systems, nervous systems, and rudimentary brains, all in the blink of a geological eye. Life diversified in the ocean, before spreading to land – first mosses and ferns, then seeded plants and the first amphibians, and then reptiles. This coincided with the end of snowball phases. Earth has not descended into a bleak, almost lifeless deep freeze since the arrival of more complex life – and this fact may be no coincidence. Instead, three milder icehouses have punctuated the dominant hothouse. As plants colonized land, life grew to assume an even stronger role in regulating carbon, nitrogen, and water cycles. This created new dynamics, harmonies, and rhythms in Earth's life support system.

Almost all life on Earth lives in a narrow band, from approximately 500 metres (1,640 feet) below the ocean floor to about 11 kilometres (6.8 miles) up in the atmosphere. Laid flat, you could cycle it in less than an hour. This is Earth's biosphere – the region where life exists. It is no wonder, given the wisp-thin layer we inhabit, that an asteroid could wipe out so much life. This is what happened 66 million years ago, thereby ending the reign of non-avian dinosaurs.

The asteroid strike was the last of the five mass extinctions that hit the reset button for life during Earth 4.0.[8] Each time, a cataclysmic event wiped out most but not all life. Mass extinctions

---

8 In addition to the five mass extinctions, there have been about 24 extinction events that do not quite make it into the "mass" category, not least the enigmatic Paleocene–Eocene Thermal Maximum event, which occurred 55 million years ago.

were, in the majority of cases, probably due to volcanic activity or the occasional asteroid impact. They led to catastrophic changes in climate, ocean acidification, and anoxia (a lack of oxygen in the ocean). The most severe occurred 252 million years ago, and it wiped out 96 per cent of marine species and a good chunk of life on land – geological evidence shows a huge pulse of carbon dioxide, hinting at volcanic activity. We are measuring similar cataclysmic changes today, which is why many biologists say that life on Earth is at the beginning of a sixth mass extinction event.

Descent into the icehouse

The dinosaurs roamed for 150 million years before the asteroid smashed a hole in Earth. For some of this time, most land was fused together into one supercontinent known as Pangaea. But tectonic activity began ripping the supercontinent apart into the continents that are more recognizable today. New oceans opened, and India, a separate continent at the time, slammed into Asia, pushing up the Himalayas some 20 to 50 million years ago. This seems to have nudged Earth's climate out of its equilibrium. As Antarctica slid down to the South Pole, it became cut off from other land masses and surrounded by cold water, which cooled the continent. Eventually, an ice cap began forming 34 million years ago. That ice cap is now 3 kilometres (1.8 miles) thick. Earth gradually descended into an icehouse[9] – our world. We finally waved goodbye to a hothouse Earth some 5 million years ago, when temperatures were about 4°C (7.2°F) warmer than they are today. As carbon dioxide levels fell further, ice formed in the northern hemisphere all year round, and sea ice at the North Pole was hemmed in by surrounding continents, allowing a permanent blanket of ice to wax and wane with the seasons.

Earth's self-regulation system

Since the biosphere first emerged 3.8 billion years ago (Earth 2.0), the Earth system has remained remarkably stable. The temperature

---

9 Unlike the hothouse, an icehouse world has a large build-up of ice permanently at the poles. However, it is quite different from the extreme snowball Earth. We live in an icehouse world, which flickers occasionally to modestly warmer interglacials.

has not left a narrow range between the freezing point and the boiling point of water. This explains why life has flourished. But this is quite curious when you think about it. Over billions of years, the Sun's brightness has very slowly increased by about 25 per cent.[10] You might expect this to cause havoc with a planetary life support system of chemical flows. Yet the major cycles of Earth's self-regulation system – carbon, nitrogen, water, phosphorus, and oxygen – have remained finely balanced and tuned for life over very long time spans. How so?

The remarkable fact that life has persisted for so long indicates that some sort of chemical, physical, geological, or biological mechanisms slam on the brakes if the temperature bobs too far beyond a Goldilocks warmth (not too hot, not too cold). Earth's climate regulation is ingenious. Volcanoes emit carbon dioxide slowly over millions of years (almost unnoticeable over centuries). If the amount of carbon dioxide kept growing, it would push us towards a Venus-like state. Luckily, something kicks in to stop it.

Earth has at least one long-term braking mechanism that dampens change by pulling carbon dioxide out of the atmosphere: silicate weathering of rocks. This is a slow geological process that occurs over millions of years. As Earth warms up, more rain falls, because a warmer ocean releases water more easily into the atmosphere and a warmer atmosphere holds more water. High in the clouds, carbon dioxide in the atmosphere reacts with the rainwater to make rain that is very weakly acidic. When this rain falls on some types of rocks, it dissolves them bit by bit, trapping some carbon from the air. The dissolved minerals wash downstream in rivers into the ocean, where marine plankton use the minerals to make their shells. When they die, these shells sink to the ocean floor and contribute to the sediment layers, trapping the carbon for millions of years.

If Earth gets warmer, this reaction speeds up and so locks away more carbon, thereby preventing the planet from overheating. As the planet cools, the reaction slows, dampening any cooling effect. Large mountain ranges cause more rain and snow to fall, thus

---

10 The Sun brightens 8 per cent every billion years.

driving faster erosion. In all likelihood, the rise of the vast Himalayas was a major cause of our gradual descent into the icehouse.

But plants and microbes help to control the mechanism, too. Or they influence weathering, at least. Plants and microbes in soils create an acidic environment that can accelerate weathering, pulling carbon dioxide out of the air faster, and they also store carbon in their roots, trunks, leaves, and branches. The arrival of plants on land 470 million years ago led to the biggest change in carbon dioxide in the atmosphere during Earth 4.0.

So carbon dioxide is the main control for Earth's climate. And life on Earth plays some sort of role in making sure the control is not yanked too hard. Occasionally, a shock such as the Mount Toba supervolcanic eruption 75,000 years ago knocks this harmony for six, but Earth has always rebounded.[11] Yet at its extreme, snowball Earth indicates that this regulation can break down catastrophically.

Intriguingly, some researchers have shown – in computer models, at least – that the stability of Earth's life support system can be enhanced, up to a point, by greater biodiversity. In fact, Earth's rich diversity of life may currently be higher than at any point in history, because diversity and complexity tend to increase over time. If the rich diversity and complexity of life increases the resilience of Earth's life support system to shocks, we would be foolish to destroy it. But this is what we are doing.

In the 1970s, James Lovelock and Lynn Margulis introduced the Gaia hypothesis, the radical and intriguing idea that the planetary ecosystem, our tangled web of life, has evolved to stabilize Earth's life support system. That is, life influences the major cycles on Earth to maintain the habitability of Earth for life.[12] The first part of the preceding sentence has been proved conclusively, but the last part remains contentious. Despite this, we can say for certain that life and Earth's life support system evolve together. They are deeply intertwined.

---

11 This rebound happens on the very long timescale of silicate weathering, which provides further evidence that this is one of the significant mechanisms for a kind of Earth homeostasis.
12 Living organisms help to regulate four Earth system processes: carbon, oxygen, nitrogen, and phosphorus cycles. As in the human body, these feedbacks create a kind of homeostasis, like a state of equilibrium.

A warning from the past

We have already mentioned several shocks that have rocked life on Earth. Here is one more: the little-known Palaeocene–Eocene Thermal Maximum (PETM) event. Let's turn the clock back 56 million years. The dinosaurs are dead; Earth is still in the hothouse, with temperatures 8°C (14.4°F) above those of today. Almost overnight, geologically speaking, temperatures shoot up a further 5°C (9°F), causing chaos on Earth.[13] This PETM event came close to "mass extinction" status as it led to a huge die-off in the ocean, caused by ocean acidification, and many extinctions on land.

Things quickly got out of control, with carbon dioxide levels skyrocketing. Some scientists think that volcanic activity somehow triggered the release of vast quantities of fossil fuels from undersea methane hydrate beds or by igniting massive reserves of oil or coal. Billions of tonnes of carbon dioxide billowed into the atmosphere... not so different from what we are doing now. Indeed, our ocean is currently acidifying at a rate that is even faster than during the time of the PETM shock.

As Earth recovered from the PETM event, new species of mammals emerged. These included the ancestors of the horse and elk, bats, and whales, as well as the first social primates. As the climate changed rapidly, primates were forced to adapt, with evolution favouring large, cohesive social groups. Large social groups require greater brain capacity, and we will discuss the implications of this in Chapter 3.

The PETM catastrophe is likely to have paved the way for the rise of humanity, but it could not stave off the impending icehouse. Over tens of millions of years, silicate weathering took a slow and steady course, stripping carbon dioxide out of the atmosphere and locking it beneath the ocean. The continents continued to drift apart. These formidable forces cooled the planet in fits and starts.

And this is where this particular story ends. We have travelled some 4.5 billion years, through three revolutions (life, oxygen photosynthesis, and complex life), each of which kick-started a new aeon, and now we settle into the past 3 million years. We have

---

13 In reality the PETM event lasted 170,000 years.

established that there are three stable states on Earth: a hothouse, more than 4°C (7.2°F) warmer than today; an icehouse, between 5°C (9°F) cooler and 1°C (1.8°F) warmer than today; and the snowball, the deep freeze.

The question is, where do we go next?

Our knowledge of Earth's life support system has grown over the past 200 years, as geologists, biologists, oceanographers, meteorologists, and many other scientific disciplines embark on a journey that never fails to delight, inspire, and surprise. In the past few decades, we have come further faster than ever before. When we peer into the deep past, through billions of years of history, one thing stands out: the impact that humanity is having on this planet right now rivals the biggest upheavals Earth has ever witnessed in its 4.5 billion years. While much of this turmoil took millions or hundreds of millions of years to unfold, our impact has materialized in a single lifetime ... and it is accelerating.

If we want to understand our world, we need to look into the ups and downs of the next 3 million years, as the ice age cycle begins. But if we want to understand the risks we take in the future, we need to look back much deeper in time.

A SCOTTISH JANITOR AND A SERBIAN MATHEMATICIAN DISCOVER EARTH'S HAIR TRIGGER

There are three stages of scientific discovery: first people deny it is true; then they deny it is important; finally they credit the wrong person.

ALEXANDER VON HUMBOLDT
NATURALIST AND EXPLORER

With all the remarkable discoveries of the past century, from penicillin and the sequencing of DNA to gravitational waves and the structure of black holes, it is easy to overlook perhaps the most astounding and important discovery for our very survival: Earth's heartbeat. Relatively recently, Earth's vital signs – temperature, carbon dioxide, and methane – slipped into a very regular rise and fall: the cycle of ice ages. It is like looking at a healthy heartbeat on a hospital monitor.

Imagine a time machine has spun back 2.7 million years. Ahead of us, an unusual cycle of ice ages is about to begin. If we look back deep in time, we can see how we arrived at this place: the long slow slide from the hothouse towards the icehouse. Initially, the first ice sheet formed in Antarctica. Then the continents of North and South

America crunched together, snapping shut the Isthmus of Panama. This changed ocean circulation and, in turn, how heat flowed around the planet. At the North Pole, the ocean began to freeze over in winter and an ice sheet formed across the northern continents.

The geological dancing of continents introduced more instability in the Earth system. This marked the start of the Pleistocene epoch, the name geologists give to the back and forth, tick-tock of ice age cycles for 2.7 million years. As we saw in Chapter 1, deep instabilities can force evolutionary revolutions. What will these instabilities bring?

The story of how science unravelled the mystery of ice ages starts with two remarkable men: a Scottish janitor and a Serbian mathematician. It finishes with an international expedition to Antarctica in the 1990s to extract ice cores stretching back through the ice ages. These ice cores are the forensic evidence unequivocally linking the industrial revolution, which began in the United Kingdom in the 1800s, to one of the most profound geological changes in the history of the planet. Let's meet the janitor first.

Tilts and wobbles

In the 1800s, geologists reluctantly began to accept the evidence that large parts of Europe, Asia, and North America were once covered in deep sheets of ice. But no one could find a convincing mechanism that would cause Earth to both drop into and pull out of ice ages. That changed in 1864, when James Croll published a remarkable analysis linking slight shifts in Earth's orbit around the Sun to ice ages.

Croll left school at the age of 13 and had a colourful career, first working as a labourer before becoming a tea merchant, then a hotel manager, and then an insurance agent, before finally joining Andersonian University, Glasgow, in 1859, as a janitor. Croll's powers of persuasion were first rate. We know this because he cajoled his brother into doing most of his caretaking work, leaving Croll more time to exercise his mind.

With no formal academic training, Croll used the university library to teach himself astronomy, among many other scientific disciplines, before he published his masterful paper, titled "On the physical cause of the change of climate during geological epochs", in 1864. In the paper, the janitor listed the most fashionable theories for the cause of ice ages and then dismissed them, one by one. Based on detailed calculations, the paper concluded that ice ages were, in fact, linked directly to Earth's orbit around the Sun, as it shifted from a more circular shape to elliptical. At the time, this was an Earth-shattering idea. It gained a lot of attention initially, but then remained in the shadows until a Serbian scientist, Milutin Milankovitch, who also had no formal education in the physics of the solar system, developed Croll's ideas to a monumental conclusion. (Croll has been largely forgotten by history, and in Earth system science it is Milankovitch who gets all the credit.)

Croll's rough estimate did not have the precision needed to convince everyone. After all, Earth's orbit is not straightforward. All the planets in the solar system orbit at different rates. Speeding around on the inside, Earth overtakes its neighbour, Jupiter, but when it does so it feels a little tug from the gas giant's gravitational field. Furthermore, Earth bulges a bit at the equator. The Sun and the other planets exert a little more force on this extra mass, thus nudging Earth's orbit. These complications, though seemingly minor in the grand celestial scheme of things, clearly matter. Combined, the circular to elliptical shift, the small wobbles as Earth turns, and the variations in Earth's tilt over time all affect its orbit in predictable, regular ways. Milankovitch set about crunching the fiendishly complicated numbers. What he discovered changed how we think about our planet.

In 1941, after some three decades of research, Milankovitch published his findings in a book titled *Canon of Insolation of the Earth and Its Application to the Problem of the Ice Ages*. With uncanny precision, it provided a link between ice ages and relatively slight changes in celestial motion. For almost 3 million years, Earth had shifted in and out of ice ages. At first, ice ages were frequent – occurring every 41,000 years or so – but relatively mild. Then, about 1 million years ago, they became less frequent – every 100,000 years or so – and ferocious. The warm interludes that occur as the ice recedes to the highest latitudes – the so-called interglacials – can last anywhere from 10,000 to 30,000 years. We have been basking in the most recent one for 11,700 years or so.

Earth's tilts and wobbles do not really affect how much energy from the Sun hits the planet, but they make a massive difference to where the Sun's energy falls and in what season. Originally, researchers thought ice ages probably started with several harsh winters. However, the Russian-German meteorologist and climatologist Wladimir Köppen worked out that it is summers that tip the balance. If one summer in the northern hemisphere is too cool to melt all the snow, then snow remains on the ground. The white surface reflects heat back to space, keeping the area cool and encouraging more snow to build up, thus creating ever more white surface. The next summer may see even more snow remaining. In this way, a small change can set off a chain reaction. Like a gun that will fire at the slightest pressure, Earth has a hair trigger in its current configuration of continents. The slightest provocation can send the system spinning and wheeling as sleeping giants, such as ice sheets and ocean circulations, awake and stretch. In later chapters, we will look at the risks of waking the gigantic components of the Earth system that we take for granted and assume, wrongly, are too big to fail.

So why hadn't celestial mechanics triggered an ice age cycle before? The levels of carbon dioxide in the atmosphere were too high. About 3 million years ago, carbon dioxide levels fell to about 350 parts per million (ppm). This was a critical tipping point, because it caused more ice to build up over North America and Scandinavia. In 1988, Earth passed this critical 350 ppm point once more, but in the opposite direction. That same year, NASA scientist James Hansen testified before the US Congress, stating that humans were warming the planet. And today we may have loaded so much carbon dioxide into the atmosphere that we are unlikely to transition into an ice age ever again.

Bubbles in the ice

One outstanding question remained: why did the pushes and pulls on Earth from Jupiter and the Sun cause such large swings in temperature on Earth? As Earth pulls out of an ice age, the temperature rises far more than we would expect. Small changes must set off other chain reactions, causing some sort of cascade that drives temperature up or down.

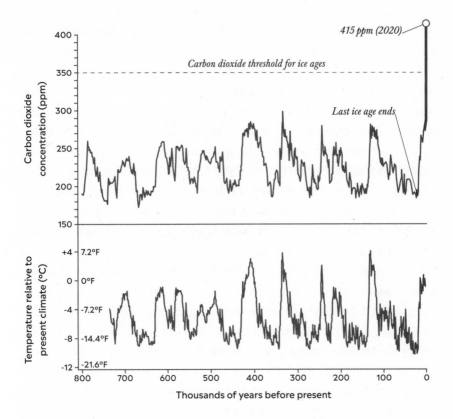

Ice age cycles

At the end of the 20th century, the final piece of the ice age puzzle slid into place. In the 1990s, international teams of scientists carted drilling equipment to Antarctica. They were not searching for oil or other minerals, but for something that may ultimately turn out to be more valuable for our long-term survival as a species: bubbles trapped in ancient ice. Pockets of air in these ice cores tell the story of the atmosphere across the past eight ice age cycles, stretching back 800,000 years. The amount of carbon dioxide and methane, two greenhouse gases, contained in the bubbles gives critical information about the Earth system in this volatile state.

The records these scientists uncovered are nothing less than Earth's heartbeat. They show us how Earth descends slowly into an ice age, but emerges rapidly. We see a triple pulse: global average temperature, carbon dioxide, and methane (temperature and carbon dioxide are shown on the preceding page). The movements of the planets are enough to nudge Earth into and out of ice ages, which sets off a cascade that amplifies the impact. But what is the mechanism to drive up greenhouse gases? A cool ocean absorbs more carbon dioxide from the atmosphere, cooling the planet further (cold water holds more carbon dioxide than warm). In reverse, as the celestial wheels spin, ice retreats and the ocean warms and releases carbon dioxide into the atmosphere, thereby causing more warmth. All this carbon dioxide and heat leads the ice to melt faster, thus releasing more carbon from the ocean and permafrost, and so on, in a self-amplifying cycle, or positive feedback loop, until it runs out of steam.

Every interglacial period is slightly different. Some are a little cooler, some a little warmer. Some a little longer, some quite short. But the instabilities all lie within tight temperature boundary conditions. No interglacial temperature has risen more than 2°C (3.6°F) above the average temperature of the past 10 millennia – our current interglacial, the Holocene – and ice ages only dip 5°C (9°F) below the Holocene average. What we can conclude is that Earth is extremely sensitive to even very small changes in greenhouse gases. We now know that Earth has a hair trigger, where the slightest thing can set it off.

The ice cores are a stark reminder of how inherently unstable the past 2.7 million years have been, how the stability of the past 10,000 years is easy to take for granted, and how much we have changed the planet in the past century. Earth's heartbeat has been disrupted by an unprecedented pulse of greenhouse gases in the past few decades, and it now looks more like a cardiac arrest. For about 3 million years, carbon dioxide levels in the atmosphere dropped as low as 170 ppm (parts per million) during ice ages and rose as high as about 280 ppm during warm interglacials. That was it: never higher or lower. Decades ago, carbon dioxide levels shot past 350 ppm and passed 415 ppm in 2019, rising at 2 or 3 ppm a year. Global average surface temperature is now 1.1°C (2°F) above pre-industrial levels. With temperatures shooting up 0.2°C (0.36°F) a decade, without drastic action the world will cross the 2°C (3.6°F) boundary – the absolute limit of temperature in the past 2.7 million years – within a few decades. We are already stomping beyond the maximum global average temperature on Earth since we left the last ice age.

Did these unusual climate pendulum swings back and forth over the past 2.7 million years affect the course of evolution on Earth? It seems so. In the next chapter, we will look again at these cycles, but this time through the lens of human evolution.

A "WISE MAN" ARRIVES

Far out in the uncharted backwaters of the
unfashionable end of the western spiral arm of
the galaxy lies a small unregarded yellow sun.
Orbiting this at a distance of roughly ninety-two
million miles is an utterly insignificant little blue
green planet whose ape-descended life forms
are so amazingly primitive that they still think
digital watches are a pretty neat idea.

DOUGLAS ADAMS
THE HITCHHIKER'S GUIDE TO THE GALAXY, 1979

We have come a long way. But how close are we to our distant
ancestors? Closer than you might think. We are only 10 generations
away from a world without an industrial revolution. Only 300
generations from the first civilizations of Mesopotamia in modern-
day Iraq and 500 from the first farmers in the Middle East. Only
3,500 generations from an evolutionary bottleneck, when the
human population collapsed to around 10,000 reproducing pairs.
And only 10,000 generations from our earliest ancestors in Africa,
where we appeared about 200,000 years ago.[14]

Swedish biologist Carl Linnaeus was the first to class modern
humans as *Homo sapiens.* The first part, *Homo* or "man", refers to our
genus[15] and the second part, *sapiens* or "wise", is the species. Do we

14 Based on the average woman having her first child at 20 years of age. A 20-year-
old is three generations from the Holocene, which petered out around the 1950s.
15 "Genus" sits between "family" and "species" in taxonomic ranking.

live up to this grand name? As a species, we are certainly clever and resourceful. We have accumulated more knowledge about our planet than any other. But are we wise? An alien observing Earth from a distant exoplanet might view the disruption of Earth's life support system as strong evidence for a lack of wisdom among its inhabitants. Perhaps one day, when the rate of change of Earth's life support system begins to stabilize, such an observer might note that some form of wisdom has finally arrived. This may be possible this century, and even as early as 2050, but to succeed we need to start the turnaround this decade. That story will come later, though, in Act III. For now, we want to take you back to the beginning.

The early days

The cartoon of a knuckle-walking chimpanzee evolving smoothly to an upright human (now often amended to include a person crouched over a computer) is incorrect. Early human evolution was more like a bush branching out in many directions. Scientists have identified about 31 hominin species, including modern humans, extinct human species, and our immediate relatives, which evolved over the past 5 million years in Africa or Eurasia. During two periods, at least six species coexisted, allowing for complex interbreeding that helped shape who we are today.[16] Now, just a single species of human is left: us.

Early human evolution seems to have emerged over four stages. First, our ancestors began walking upright about 5 to 10 million years ago. Evidence indicates that they were already using simple tools about 3.3 million years ago. The second stage begins about 1.8 million years ago, when a new branch grew into *Homo erectus*. This species stood more upright than its predecessors. *Homo erectus* had a slightly different body shape, and child development was noticeably delayed. But perhaps its most distinguished feature was a much larger brain, which continued to expand for half a million years or more.[17] *Homo erectus* controlled fire, which made cooking

---

[16] Today, non-African populations outside Oceania carry between 1.8 and 2.6 per cent Neanderthal DNA.

[17] Between our first ancestors and *Homo sapiens*, our brains tripled in size. Most growth occurred during the time of *Homo erectus*.

possible, thus providing more energy. This would have aided further brain expansion as brains devour a lot of energy.

From *Homo erectus*, another species emerged around 700,000 years ago: *Homo heidelbergensis*, the third stage. This species had a massive brain compared with its ancestors. Indeed, its volume overlaps with modern humans. It seems speech and language may have emerged at this stage, thanks to a series of genetic mutations. It is likely that *Homo heidelbergensis* is the common ancestor of both modern humans and our close relatives, the Neanderthals. Finally, *Homo sapiens* arrived on the scene around 200,000 years ago – although new evidence indicates we may have arrived 100,000 years earlier. Up until two decades ago, scientific consensus favoured a strong East Africa origin story in the Great Rift Valley. The truth may be a little more complex. Recent research shows *Homo sapiens* evolved not as a single group in one area, but rather as a set of interlinked groups living across Africa, whose connectivity altered through time as the environment changed.

Brain power

Modern humans are not just marginally smarter than other primates and our closest relatives – there is no comparison. Our huge brains endow us with impressive mental abilities, albeit at a high cost: our brains drain 20 per cent of our daily energy. This, however, is impressively efficient because our brain's power rating is about 13 watts – similar to a low emissions light bulb.

Three competing ideas explain how and why our brains evolved so rapidly. The first is the social brain hypothesis. We needed large brains in order to live in big groups and work out complex relationships and hierarchies in order to cooperate, manipulate people, and avoid exploitation.[18] This is a strong hypothesis because social complexity, particularly competition, triggers a cognitive arms race, driving a cycle of change as people or groups attempt to triumph by outsmarting each other. The second is the ecological intelligence hypothesis. Large brains allowed us to improve how we

---

18 It goes beyond living in large groups and extends to competing with other large groups.

hunted for food, created tools, and lived in a constantly changing environment. The four phases of human evolution leading to this explosion in cognitive power coincide with the major climate shifts driven by ice ages. With each environmental upheaval, we had to rely more on our brains to find ways to survive.

The third and final hypothesis to explain rapid brain expansion is cultural intelligence. This emphasizes accumulated cultural knowledge and the role of teaching and learning. This hypothesis combines elements of the first two and relies on evolutionary steps to allow language to emerge.

The earliest ice ages during this period started 2.7 million years ago and lasted about 41,000 years, with gentle transitions from ice age to interglacials and relatively mild ice ages. These may have had little impact on equatorial Africa or the Great Rift Valley, where many of our ancestors roamed. It was not until 1 million years ago that the frequency and severity changed to 100,000-year cycles. Ice built up gradually, followed by a "sudden" collapse over 4,000 years, driving Earth into warmer interglacials. Climate patterns changed in Africa, with mega droughts alternating with rainy phases. With a changing climate and a shift from more dense forests to grasslands and deserts, early humans had to adapt. These conditions may have favoured either the smartest or those best able to cooperate within groups and compete between groups. Having all these traits was clearly the winning formula. About 100,000 years ago, our brain size changed markedly; this appears to coincide with the appearance of fully modern language.

Recently, researchers created a computer model of how our brains might evolve under similar social and ecological pressures to those probably faced by our ancestors in the Great Rift Valley. They found that the ecological challenges of extracting energy from a rapidly changing environment may have been the dominant driver of rapid brain evolution, with a supporting role for social competition between groups. This is backed up by independent evidence that indicates that while *Homo sapiens* and Neanderthal brains are quite similar, by 100,000 years ago, the brains of *Homo sapiens* had become a little more bulbous and round. These shape shifts evolved along a trajectory towards greater hand–eye coordination, complex tool use, self-awareness, long-term memory, and numerical processing.

This evolutionary path continued until 35,000 years ago, when the *Homo sapiens* brain became comparable with a fully modern human brain. In 2020, the psychologist Simon Baron-Cohen proposed a new theory to explain the significance of these changes. He argues that evolution during this period led to a unique mechanism for analysing patterns and systemizing information. This mechanism created a deceptively simple algorithm that Baron-Cohen describes as "IF, AND, THEN." (IF I plant a seed AND it rains THEN a plant will grow.) This algorithm is the mother of all invention. It opens up a world of innovation and creativity, from agriculture to carpentry, from science and mathematics to art and literature.

But why did Neanderthals, with very similar brains, not evolve along this route? Other factors may be at play. A close look at the skull evolution of *Homo sapiens* over 200,000 years shows that our prominent brows, which we shared with Neanderthals, gradually reduced over this critical time. Our faces become more rounded. Strong features that tend to define maleness decreased. These changes might be explained by lower testosterone levels. Chimpanzees have high testosterone levels, live in small groups, and settle disagreements with explosive violence. Bonobos, on the other hand, who evolved just across the Congo River from chimpanzees, have lower testosterone levels and live in large, more peaceful groups. Lower levels of testosterone do not require a major genetic mutation, but simply a little genetic diversity. What if females deliberately chose less aggressive males? Or aggressive males became increasingly shunned by tribes, allowing for more peaceful, stable, and significantly larger societies. These strategies have been called "survival of the friendliest". All this leads to an interesting conclusion: we humans domesticated ourselves.

The evolutionary twists and turns that produced a brain like ours throw up a vexing conundrum. Our brains are hard-wired and finely tuned for reasoning. You would think this might allow us to make better decisions. But as individuals and as societies, we are often hopeless at using facts and evidence. This is a particular challenge for us as a globally interconnected species that is running out of time to address a complex planetary emergency. As language and cooperation co-evolved, a particularly powerful trait emerged: persuasion – our ability to coerce, cajole, and manipulate. We are

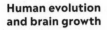

Human evolution and brain growth

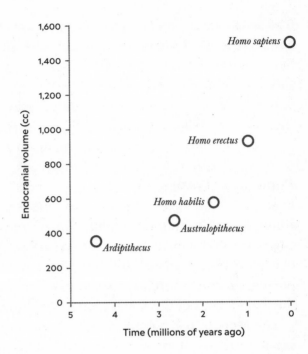

undoubtedly masters at this game. But, the power of persuasion is not necessarily enhanced by facts, much to the chagrin of scientists. This can be explained by how and why persuasion as a trait emerged in the first place. As Adrian Bardon, author of *The Truth About Denial* (2019), puts it, "Our ancestors evolved in small groups, where cooperation and persuasion had at least as much to do with reproductive success as holding accurate factual beliefs about the world." Survival in a tribe, at least until reproductive success, meant fitting in with the group's belief system. An instinctive bias that favours our group – and its world view – over others is ingrained in human psychology and DNA.

To summarize, the cognitive revolution possibly included the evolution of two new ways in which our brains operate. First, a unique capacity for systemization: the IF, AND, THEN innovation engine. And, second, an ability to cooperate, empathize, teach, and even engage in deception that has allowed very large stable groups, tribes, societies, and civilizations to emerge. But here's the crucial point. As Baron-Cohen writes in *The Pattern Seekers* (2020), "When we step back and look at these two remarkable brain mechanisms that set modern humans along a path diverging from all other animal species, we discover remarkable diversity in the population." Some people are brilliant at systemizing information but struggle to socialize in groups. For others, it is the opposite. Somehow, the tensions between these two cognitive superpowers define our successes and failures to this day. This is a particularly important insight as we now grapple with managing an Earth system.

Population bottleneck

Even with our formidable cognitive power, it seems we came within a hair's breadth of annihilation. At a critical moment in our cognitive development, some 70,000 years ago, a catastrophe befell our ancestors. Genetic analysis shows that the *Homo sapiens* population crashed to just 10,000 breeding pairs.[19] You and I are the descendants of this small group. What could have happened?

The Mount Toba volcanic eruption approximately 75,000 years ago was the largest in 2.5 million years and is likely to have led to a catastrophic global volcanic winter that lasted for decades, followed by centuries of lower temperatures and drier weather. Until recently, the Toba volcanic eruption has been the leading contender as the likely cause of the population crash. This is now disputed. A recent study of conditions in Africa after the super-eruption concludes that "humans in this region thrived through the Toba event and the ensuing full glacial conditions". So, for now at least, the reason for the bottleneck remains a mystery.

What stopped our brains expanding further? Of course, there are some physical limitations, most notably female pelvis size limiting foetal brain size and the energy consumption required for an ever-increasing brain size. But also, another revolution was just around the corner: the agricultural revolution. As the ice from the last ice age receded to the far north, we changed the way we found our food. This had quite an interesting influence on our brain development, but not in the way you might expect.

---

19 Our genetic diversity is extremely low compared with other primates like chimpanzees, indicating at least one population bottleneck.

THE GOLDILOCKS EPOCH

I am quite literally from another age.
I was born during the Holocene, the name
given to the 12,000-year period of climatic
stability that allowed humans to settle, farm
and create civilizations. Now in the space
of one human lifetime, indeed in the space
of my lifetime, all that has changed. The
Holocene has ended. The Garden of Eden
is no more. We have changed the world
so much that scientists say we are now in
a new geological age – the Anthropocene –
the Age of Humans.

DAVID ATTENBOROUGH
WORLD ECONOMIC FORUM, DAVOS, 2019

The history of humanity is really the history of the Holocene. It is on this stable stage that all of humanity's dramas have unfolded and have been recorded.

The story of the Holocene has a very definite beginning and a very definite end. It began 12,000 years ago and it ended in the 1950s. After a deep freeze lasting 100,000 years, the ice age waned under the influence of celestial mechanics and the planet's internal dynamics. Earth did not slip smoothly into the new epoch. *Homo sapiens* endured a bumpy ride at first, as warming led the ice sheets to collapse, causing immense flooding. This briefly plunged the planet back into ice age-like conditions, when the cold fresh water caused ocean circulations to grind to a crawl.

Eventually, Earth settled into what geologists call the Holocene (from the Greek, meaning "new whole") epoch,[20] one of the regular warm episodes between ice ages. We call it the Garden of Eden or Goldilocks epoch, because it was not too hot and not too cold. The defining features of the Holocene were its unusually stable climate, even compared with other warm episodes between ice ages, and the vulnerability of this stability. As discussed in Chapter 2, Earth has a hair trigger and anything could set it off. We can calculate the predictable shifts in Earth's tilt and wobble and work out that our Goldilocks epoch could have been expected to hang around a further 50,000 years if the trigger had not been squeezed so hard. In fact, our conclusion is that it is the benign environmental conditions in the Holocene that enabled us to develop from a few scattered hunters and gatherers to the hyper-connected globalized society we live in today. But our future is no longer in the Holocene.

The stories of the Holocene

The Holocene is the story of how one species came to dominate a planet and overwhelm the life support systems that keep that planet in its uncanny, relatively stable state. But it cannot be told as a single narrative; it is many. The Holocene is the story of an entangled society and environment, particularly the climate and the biosphere. It is the story of innovation and ingenuity. It is the story of cooperation and coercion, hierarchies and power, but also the story of how networks disrupt these hierarchies. The Holocene is the story of growth, particularly in its final chapter as the industrial revolution kicks in. Growth leads to complexity, and complexity leads to unexpected emergent behaviour.[21]

---

20 In geological jargon, epochs are significantly less Earth-shattering than aeons.
21 People often confuse "complicated" and "complex". A complicated task, for example, is building a rocket to take men and women to the Moon. A complex task is raising a child: whatever the inputs, the output (one adult) is quite uncertain. An example of emergent behaviour in a complex system is consciousness. It is not at all obvious from looking at a brain's neurons and synapses that some sort of grand awareness can materialize.

SLEEPING GIANTS: ANTARCTICA

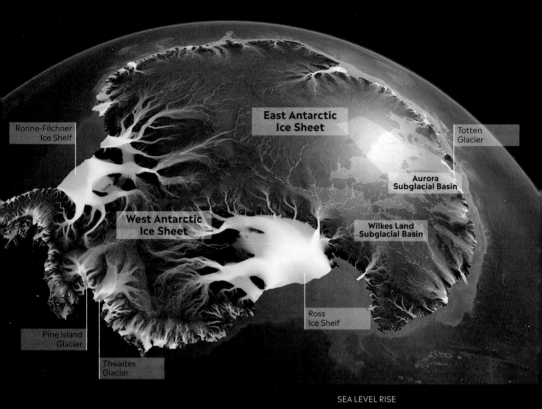

Ronne-Filchner
Ice Shelf

East Antarctic
Ice Sheet

Totten
Glacier

Aurora
Subglacial Basin

West Antarctic
Ice Sheet

Wilkes Land
Subglacial Basin

Pine Island
Glacier

Thwaites
Glacier

Ross
Ice Shelf

ICE VELOCITY

m/year

| 0 | 200 | 400 | 600 | 800 | ≥1000 |

| 0 | 600 | 1200 | 1800 | 2400 | ≥3000 |

ft/year

SEA LEVEL RISE

+9m (+30ft)
Aurora Subglacial Basin

+3.3m (+11ft)
West Antarctic Ice Sheet

+1.2m (+4ft)
Pine Island Glacier +
Thwaites Glacier

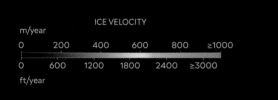

Antarctica holds enough ice to raise sea level by 60 metres (200 feet).
About one-third of the Antarctic ice sheet rests on rock below sea level. These areas are
particularly vulnerable to climate change. They are exposed to warm ocean water that
can seep underneath the ice, melting it rapidly from below and accelerating ice loss.
The ice sheet has lost trillions of tonnes of ice in recent decades. Major areas (Pine
Island Glacier, Thwaites Glacier, and parts of Wilkes Land) are already showing signs
of destabilization. A collapse of Pine Island Glacier and Thwaites Glacier would lead to
a rise in global sea level of 1.2 metres (4 feet). In total, the vulnerable area of the West
Antarctic Ice Sheet holds enough water to raise sea level by 3.3 metres (11 feet), and its
disintegration is likely to be accelerated by the collapse of Pine Island Glacier and
Thwaites Glacier. In East Antarctica's Wilkes Land, the vulnerable Aurora Subglacial
Basin holds enough water to raise sea level by 9 metres (30 feet).

RUPTURE WITH THE PAST

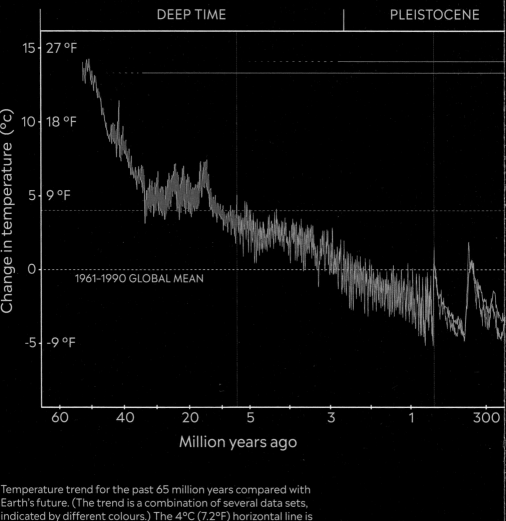

Temperature trend for the past 65 million years compared with Earth's future. (The trend is a combination of several data sets, indicated by different colours.) The 4°C (7.2°F) horizontal line is based on a 1961–1990 baseline for global average temperature.

Past: Hothouse Earth conditions prevail during this period, with a heat spike (the Paleocene-Eocene Thermal Maximum) 55 million years ago. Greenhouse gas levels fall gradually, cooling the planet until permanent ice can form in the northern hemisphere. A new climate regime begins with two stable states: ice ages and warm periods known as interglacials.

Future: Decisions made now will have a significant influence on whether or not Earth restabilizes within a cooler Holocene-like climate or lurches towards more hothouse-like conditions.

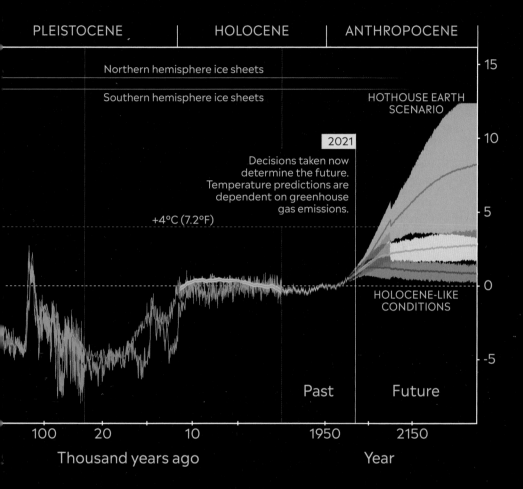

PLEISTOCENE   HOLOCENE   ANTHROPOCENE

Northern hemisphere ice sheets

Southern hemisphere ice sheets

HOTHOUSE EARTH
SCENARIO

2021

Decisions taken now
determine the future.
Temperature predictions are
dependent on greenhouse
gas emissions.

+4°C (7.2°F)

15

10

5

0

-5

HOLOCENE-LIKE
CONDITIONS

Past   Future

100   20   10   1950   2150

Thousand years ago   Year

THE EARTHSHOT

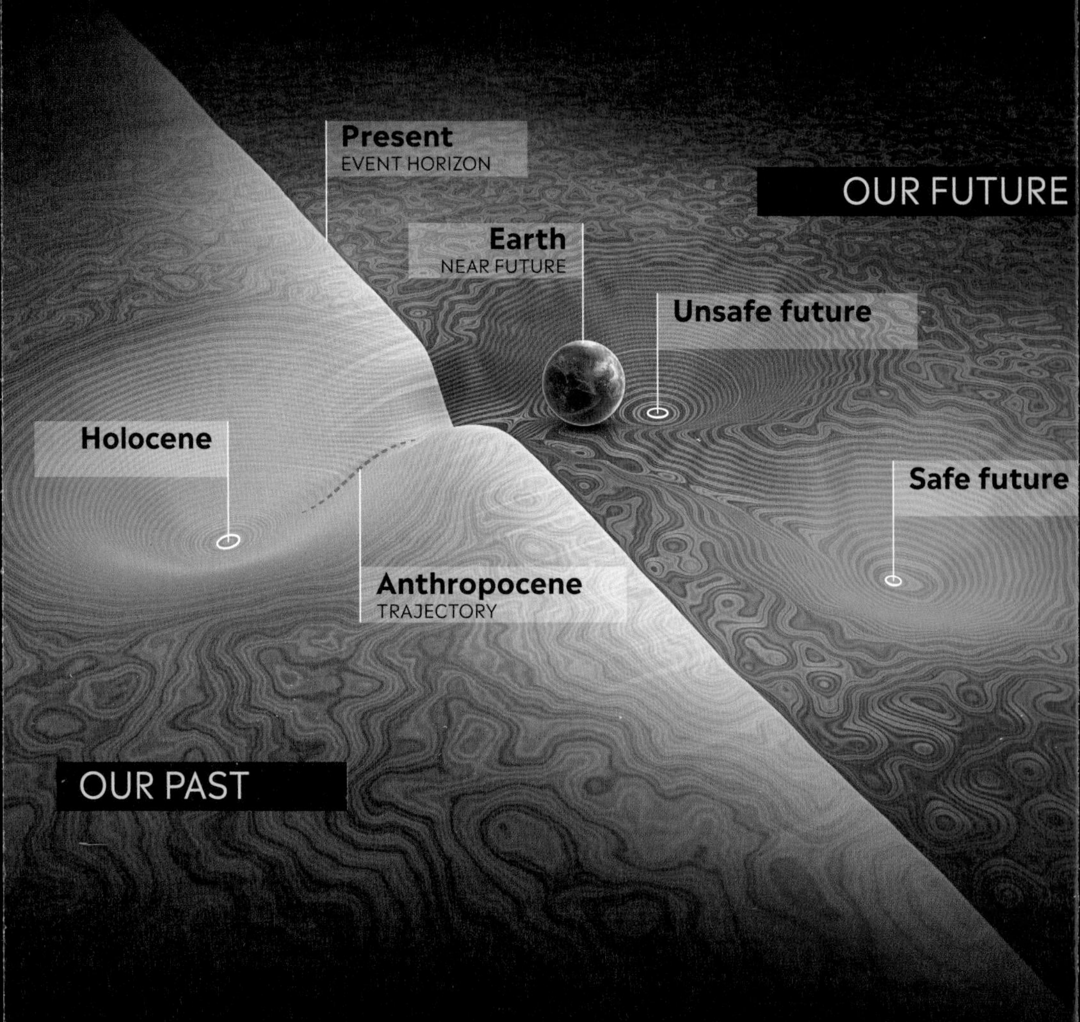

Present
EVENT HORIZON

Earth
NEAR FUTURE

Unsafe future

Holocene

Safe future

Anthropocene
TRAJECTORY

OUR PAST

Earth has slipped out of the Holocene stability regime. Our Earthshot mission is to locate and navigate towards a safe, stable state for Earth.

A4

We cannot, here, tell the complete story of our journey through the Holocene. What we want to emphasize is the remarkable climatic calmness and environmental richness that has allowed a technology-advanced civilization to flourish. Here, we want to provide enough information to grasp what we are losing, what is at stake, and why it is so difficult for complex societies to act. We also want to offer hope. War, famine, and disease once loomed large over all our lives, but, comparatively speaking, they have now all but faded away. The rise of democracy and women's rights, as well as the end of slavery, show that societies can do the right thing ... but sometimes it feels as if this only happens once they have run out of all other alternatives. We should not let the progress already made lull us into a false sense of security. Despite constant scientific warnings, in 2020 COVID-19 somehow caught the world off guard.

The story of an entangled society and environment

By the start of the Holocene, *Homo sapiens* populations had reached all continents except Antarctica – globalization began early on. But the turbulent climate meant there was little chance of agriculture emerging. For that, there needs to be some likelihood, at least, that the coming year will be as predictable as the previous one.

Eventually, the climate settled into a rolling, pleasant rhythm. This predictability has continued for so long – thousands of years – that we have absolutely taken it for granted. None of us have ever lived otherwise. Exactly how mellow has only been established with a high degree of certainty in the past two decades. Global average temperature bobbed up or down by just 1°C (1.8°F) during the Holocene. Earth's most important greenhouse gas controlling the thermostat, carbon dioxide, hovered around 280 parts per million (ppm) and did not budge until the industrial revolution.

However, the climate did not stabilize simultaneously around the world as we entered the Holocene, and so farming emerged in pockets and the domestication of plants and animals arose as environmental conditions improved regionally. Scientists have found the first signs of agriculture 11,000 years ago in Mesopotamia, the once fertile lands between the Tigris and Euphrates rivers in modern-day Iraq. Around 10,000 years ago, agriculture emerged independently in China and Central America. Then, about 8,000 years ago, agriculture began in India, Africa, and North America, as well as in parts of the Andes, in South America, as the climate became more clement. Our ancestors first domesticated, in order, dogs, wheat, barley, lentils, cattle, pigs, chickpeas, and cats.[22]

Domesticating animals and plants occurred independently on different continents around the world. This is proof that it was the Goldilocks conditions of the Holocene – giving reliable rainy seasons and warm growing seasons – that enabled us to carry out the most important revolution of all time: inventing agriculture. It was from here that everything we call "modern societies" evolved.

The agricultural revolution had a profound impact on our planet. At the start of the Holocene, about 6 trillion trees grew on Earth. Today, there are only about half that amount. Although some argue that as agriculture spread globally in its first few millennia farming practices led to widespread deforestation, which in turn affected greenhouse gases in the atmosphere, the impact was probably within natural boundaries. It is only much later that deforestation started interfering with the functioning of Earth's life support system.

Sceptics like to question whether climate change is a big deal. They point out that the Romans grew vineyards in Britain during the "medieval warm period" and that Londoners skated on the frozen Thames during the "Little Ice Age". These events were regional bubbles, though, affecting only certain parts of the globe. They were linked to Earth's natural climate cycles and occasional volcanic eruptions. But globally, the average temperature never crossed the narrow boundary of plus or minus 1°C (1.8°F). Sceptics also say the climate has always changed: look at the ice ages or hothouse Earth condition. Yes. This should alarm us, because Earth can be extremely sensitive to disturbances, yet the climate barely changed during the Holocene, the time when civilization emerged.

Not all of Earth is equally suitable for vast networked civilizations. The frozen tundras and forests of the north are cold, bleak, and

---

22 This is based on the first identification of changes to the shape or morphology of these species as a result of human interference. Some form of farming will have emerged some time before this, because wild plant and animal species will have been managed before morphological changes were introduced. Some cats may dispute whether cats have, indeed, been domesticated.

inhospitable. The tropics are blanketed in dense rainforests and tropical diseases are rife. Indeed, only a narrow band around Earth – known as "the lucky latitudes" – has seen the most advanced political, technological, and economic leaps and bounds. It is here where the largest populations have flourished.

Throughout the Holocene, we relied on a stable environment for our well-being. We told stories and adopted world views that acknowledged the need for environmental stewardship to protect that stability. Those world views informed how we farmed, harvested, and lived our lives. Small island communities were more aware of this than others. Even so, these stories did not prevent disaster in precarious isolated locations. The statue-building Rapa Nui society of Easter Island in the Pacific Ocean collapsed in the 1600s or early 1700s, possibly as a result of chopping down the island's trees, which held topsoil in place. While global trade now provides more resilience for once-isolated societies, it can also make the whole network more prone to shocks. Moreover, societies are increasingly disconnected from nature, which augments our vulnerability. Look no further than the 2020 pandemic for evidence of this.

The story of innovation and ingenuity

The Sumerians of Mesopotamia probably win the prize for first civilization. As farms and communities grew more complex, this created surplus food. By 7,000 years ago, several cities had emerged on the plains of Mesopotamia, some eventually with populations of 80,000 people. It is here we find the earliest writing: cuneiform.[23] The Sumerians used cuneiform for accounting, to track trades and take stock. Later, it was used for governance, law, documenting historical events, and storytelling.[24] Writing was a foundational

---

[23] Incredibly, up to 2 million cuneiform tablets have been excavated, although not all have been translated. Clay is an impressively resilient medium.
[24] The first written story is the Epic of Gilgamesh, an account of a hero/king, which includes the first reference to a "Great Flood". Great Floods appear in Norse, Celtic, Chinese, and Christian mythologies, and there is speculation as to whether these written accounts are based on oral stories, passed down through generations, and relate to the floods that must have inundated early human settlements as ice sheets in the north collapsed catastrophically in the rough ride into the Holocene.

innovation for organized societies, allowing expansion of civilizations, empires, and their attendant bureaucracies.

Like language before it, writing makes all other innovation both possible and efficient – no one needs to reinvent the wheel if someone has written down the instructions. Both writing and language allow culture and knowledge to accumulate. But in a world shaped by great empires and religious thinking, access to writing and reading became an elite privilege that was used to exert control, thus stifling ideas and so technological and social innovation for millennia. It took 6,000 years for that to change, when Johannes Gutenberg introduced another foundational innovation: printing. This was a decisive point in the Holocene. The British historian Niall Ferguson notes that the subsequent diffusion of printing presses throughout Europe in the 1500s accounted for 18 to 80 per cent of urban growth. Printing allowed people to disseminate fundamental knowledge for well-functioning economies: how to brew beer, for example. It also allowed radical new ideas to diffuse through societies, sowing the seeds of the Enlightenment and scientific reasoning.

The story of cooperation and coercion, hierarchies and power
In our evolutionary history, few developments were as critical in shaping us into modern humans as our ability to cooperate. Irrigation, one of the most important innovations in the early Holocene, is an engineering challenge. But it is also a social challenge. Irrigation channels meant that farmers could plant more seeds and provide some insurance against drought. A lone farmer would struggle to create such a system. Instead, people needed to come together and cooperate as larger groups. Irrigation required collaboration to such a great extent that some academics have proposed the systems as important catalysts for social innovation, centralized management of resources, and the origin of the state.

The earliest states relied on irrigation because it allowed agricultural intensification and, in turn, surpluses – essential to support those living in the burgeoning towns and cities. Eventually, similar technologies brought water to cities and carried waste away. Recently, water usage has skyrocketed. About 70 per cent of this increase is for irrigation. Since 1900, water usage increased six-fold, with most change happening since the 1950s as the green revolution – a huge global investment in technologies like tractors and fertilizers – brought food security to billions more people. It now looks as if it may be plateauing – mainly because there are very few rivers left that can be dammed or siphoned – but the World Bank estimates that by 2050, agriculture may need to draw an additional 15 per cent of water to feed a growing population. Where will this come from?

As cities grew more powerful, empires emerged, with power tightly concentrated at the top in a patriarchy of chiefs, kings, and emperors. Technological innovation in the hands of hierarchical states focused on war machines and architectural feats that symbolize power, such as the Egyptian or Maya pyramids or the cathedrals of Europe. In this world, power and energy came from trees, slaves, horses, and oxen, and information flowed only downwards from upon high, if at all. Major engineering projects, such as the massive irrigation schemes in Mesopotamia, could be ordered, but subsequently protecting crops from saltwater intrusion seemed beyond the power of these vast empires.

The story of how networks disrupt hierarchies

The story of empires is well known because it is documented by the empires themselves in their bureaucracies. The story of networks is less known[25] and more intriguing. Ferguson argues in his book *The Square and the Tower: Networks and Power, from the Freemasons to Facebook* (2018) that "innovations have tended to come from networks more than from hierarchies", particularly since the invention of the printing press.

The Reformation in Europe, which created a huge chasm between the Roman Catholic Church and the political elite in northern Europe, started from a small informal network able to print its ideas and spread them at speed across Europe. More significantly, the Enlightenment movement – which was the foundation of the scientific revolution, the industrial revolution, and the philosophy, law, and economic thinking that created modern democratic, liberal

---

25 This is because of their inherently ephemeral nature.

societies – emerged from informal networks of intellectuals. One of these, Adam Smith,[26] at the University of Glasgow in Scotland, argued that a free market would allocate resources more efficiently than a rigid hierarchy – an idea that became a foundation for capitalism. (In reality, a quick glance at retail giant Walmart's supply chains shows that impressive command and control is required to get goods to markets. A global economy needs both command and control hierarchies and network effects.)

But networks have one major downside: they are not easily directed towards a common objective. Ferguson says, rightly, that networks are "creative but not strategic" and are as capable of spreading bad ideas as good ideas. The digital revolution has allowed us to collect sample data to prove this. Indeed, Facebook and Twitter have finely honed algorithms to industrialize the spread of bad ideas, seemingly as part of their business model, such as fake news about climate change and vaccines. The Flat Earth Society is thriving in the digital era.

The story of growth

The US author and academic Jared Diamond famously called the agricultural revolution the worst mistake in human history. This is a provocative statement, challenging the notion that the move from hunter gatherers to farmers was progress. Certainly, there is evidence that our ancestors had to work longer hours to put food on the table and that the food may not have been as nutritious. But agriculture allowed populations to grow larger because, under the plough, a hectare of land could feed more people. Combined with a shift to a more sedentary way of life, agriculture allowed towns and then cities to grow. This had more benefits than downsides, with a growing number of people enjoying a higher quality of life.

---

26 Adam Smith invited a young James Watt to Glasgow University, where they became good friends. Watt is credited with improving the steam engine to such an extent that it would kick-start the industrial revolution, as we will see in the next chapter. You may recall that James Croll made his monumental discoveries in Glasgow, too. What made Scotland a crucible for the scientific and industrial revolutions? Education.

**Human societies and their
increasing environmental impact**

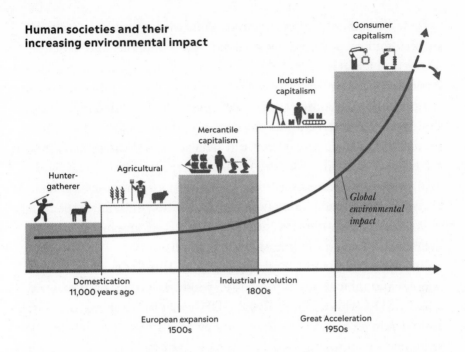

The reality is, though, that despite the onset of agriculture, empire building, urbanization, and trade networks, the global population grew modestly for much of the Holocene. Economies grew only modestly, too. This was not exponential growth but more linear and based on population size. When cities reached a certain size, the sheer density of people crammed into confined spaces, which now provides such a powerful network effect and drives new ideas and innovation, allowed pandemics to explode more efficiently. The Black Death of the 1340s killed up to 200 million people across Asia and Europe. On top of that, few incentives existed to extract natural resources and turn them into useful products on any scale. In other words, the vast empires of the past had limited desire or ability for any kind of continued or even noticeable economic growth. However, the advent of capitalism, market forces, and colonialism allowed greater resource extraction and brought riches to countries such as the United Kingdom, Spain, and Portugal. But even with the superpower of capitalism, economies did not really take off until the 1800s when coal, and later oil, extraction began providing undreamed-of energy on tap. Strangely, most economists omit energy in their analyses of growth.

In 1775, Scottish engineer James Watt worked out how to make steam engines powered by coal far more reliable and efficient. Mines and factories across the United Kingdom rapidly adopted his invention, which paved the way for railways. In the nascent textile industry, the spinning jenny accelerated wool production, flying shuttles doubled the outputs of weavers, and cotton gins made linen production at scale possible (as did slavery in the Americas). Then, electricity brought lights, telephones, radios, televisions, refrigerators, freezers, and more. The inventions of modern concrete and mass steel production allowed urbanization and better roads. It seemed as if each new invention created a springboard to greater heights and fuelled the industrial revolution.

By 1804, the world had crossed the threshold of 1 billion people and global gross domestic product (GDP) had reached about USD 1 trillion. From there, GDP began to climb rapidly. The United Kingdom was the first country to achieve the alchemy of sustained economic growth, as factories emerged in the north of the country, but this quickly spread across Europe, North America, and elsewhere. Global GDP is now about USD 86 trillion, a staggering leap over the past 215 years.

In 1800, about 90 per cent of the global population lived in rural areas. As the industrial revolution gathered pace, people swarmed to cities, which became engines of creativity and growth. Today, we are an urban species: 51 per cent of the population lives in cities, although one-third live in slums. Tokyo has the largest population, numbering a staggering 38 million inhabitants; perhaps unsurprisingly, it is the most economically productive city on Earth (followed by New York).

City metabolisms have colossal appetites for materials and energy. While people in densely populated cities tend to have lower carbon footprints, due to more efficient mass transit systems and smaller apartments to heat, the sheer numbers require global trade flows on unprecedented levels. And rapid urbanization requires steel, cement, and a colourful palette of toxic chemicals. In the 20th century, the United States used 4.6 billion tonnes (5 billion tons) of cement to build its economy. China used more in just three years between 2008 and 2010. Urbanization is accelerating. By 2050, 7 billion people will be living in cities.

While the agricultural revolution at the start of the Holocene allowed our ancestors to feed more people using a smaller area of land, by the industrial revolution farmers had reached the limit of what a hectare of land could produce. Industrialized countries were going to greater lengths to source guano, or birdshit, which is a superb fertilizer. At the start of the industrial revolution, millions of tonnes of guano were being shipped from South America to Europe every year to fertilize crops. Then two Germans, Fritz Haber and Carl Bosch, developed a way to make nitrogen fertilizers using nitrogen in air. This was one of the most significant inventions in our history, and it has allowed a staggering expansion of food production.

One thousand years ago, just 4 per cent of global ice-free, non-barren land was used for agriculture. Now, we use an area the size of South America for our crops and an area the size of Africa for our livestock, or a full 50 per cent of the habitable land surface: half Earth. But this figure would be far bigger were it not for improvements in crop yields as a result of artificial fertilizers and other technologies. Between 1860 and 2016, it is estimated that 128 million people died in famines. The causes of famine are complex and relate to trade, extreme poverty, and price volatility, as well as crop failure. Since the 1960s, major famines have all but been eradicated. Until very recently, famine was thought to be a thing of the past outside of politically unstable states. This view is altering as the climate changes and Earth reaches a saturation level and can no longer absorb the punishment from human impact.

All these developments, particularly improvements to health and agriculture, kick-started population growth. Between 1804 and 1927, the population jumped by 1 billion. It doubled again to 4 billion in 1975 – in just 48 years. At the time of writing, we stand at 7.8 billion. But population growth has slowed down dramatically as a result of progress in gender equality, particularly the increases in women's education and greater economic opportunities for women. The rate of population growth has halved since its peak in the 1960s and is now at a crawl. We are reaching "peak child", a term popularized by the late Swedish academic Hans Rosling. Peak child is when the number of children reaches the replacement rate of the population: that is, about two children for every woman.

The global population may peak at 10 to 11 billion people this century, then fall to a little below 10 billion. This is still a huge figure, obviously, but as we will see in Act III, transformation towards healthy diets and sustainable agriculture means Earth can probably support such a population.

One of the main reasons that the population is continuing to rise, albeit more slowly, is longevity. In the 120 years since 1900, life expectancy doubled globally from 35 years to over 70. As hygiene and medicine improved, more and more people were surviving at all ages. Most prominently, there was a staggering reduction in child mortality. Vaccine development, which began in 1796 with the first smallpox vaccine, was also critical. By the end of the 19th century, the first vaccines for cholera, typhoid, plague, and rabies had been developed, and antiseptics use for surgery was widespread. In 1928, Scottish scientist Alexander Fleming accidentally discovered penicillin. Then, in the 1950s, James Watson, Francis Crick, and Rosalind Franklin unravelled the secrets of DNA, ushering in another revolution in medicine.

In 1800, most people lived in extreme poverty. Today, less than 10 per cent, or 700 million people, live on less than USD 1.90 per day, the definition of extreme poverty. The number of people who are not in extreme poverty has increased more than 50-fold. The world's exit from poverty hit an inflection point around 1950 and reached a second in 2000 as population growth slowed. Of course, we should not forget that those living on USD 2 per day are still desperately poor. This is changing rapidly, though: a decade ago, only a quarter of the world population earned more than USD 10 per day; now that figure has reached a third, mainly due to phenomenal economic growth in China and also India.

While futurologists like to say technological innovation is accelerating, which is true, the nature of innovation in the past two centuries and the impact it has had on people and the planet are never likely to be repeated: electricity, antibiotics, and nitrogen production are truly game changers. The iPod less so.

Economic prosperity is a recent achievement for humanity. Growth over the past two centuries has created a modern world with less poverty, greater longevity, increased security, and greater well-being than at any point in history. When you look at the trends, most of this growth has occurred since the 1950s. This is when the industrial revolution went into overdrive. But this growth has come at a high cost. Growth based on fossil fuels and natural resource extraction is now unequivocally interfering with the stability of Earth's life support system. It is also undermining the resilience of Earth's biosphere. The stable Holocene is behind us.

Carbon dioxide levels in the atmosphere have shot up. They are now higher than at any time in at least 3 million years. But they are not just nudging over the Holocene boundary, or even the interglacial boundaries over this period; they are 50 per cent higher than at any other time in the Holocene and growing rapidly. We see similar trends for methane, another greenhouse gas, and nitrous oxide yet another. The ocean is becoming more acidic at a rate not seen in possibly 300 million years; it is losing oxygen, and currents are changing. Heatwaves are sweeping through the ocean, killing coral. We have lost a full half of the Great Barrier Reef in the past decade. We have made a hole in the ozone layer. We have changed the global water cycle, but also the cycles for carbon, phosphorus, nitrogen, and more. Humans move more rock than all natural processes. All this is driving the sixth mass extinction of life on Earth.

During the Holocene, a typical adult's food intake and energy needs translated to a power rating of about 90 watts a day. The average American now uses 11,000 watts every day – the energy needs of about a dozen elephants.[27] The scale of the shift in energy consumption is barely comprehensible.

What caused the Great Acceleration?
Over the past few decades, scientists from many disciplines have joined forces to put together the most comprehensive picture of what the hell just happened. Initially, they considered the start of the industrial revolution as the moment when humanity became a planetary force. But that view has changed recently. There is now overwhelming evidence that it was really the 1950s when humanity began to overpower Earth's life support system.

---

27 If the global population consumed energy at this scale, it would be the equivalent of the energy needs of more than 90 billion elephants roaming Earth.

After the Second World War, the conditions were ripe for one of the most profound events not just in human history, but in the history of our planet. The start of this period has been referred to as the Golden Age of Capitalism or *Les Trente Glorieuses* (The Glorious Thirty). It is more accurately called the Great Acceleration, and it continues to this day. The data behind the Great Acceleration led to the profound realization that Earth has left the Holocene, our Goldilocks era. In a single human lifetime, Earth has crossed into an altogether less certain epoch: the Anthropocene.

Some academics say the Anthropocene is misnamed; it should be the Capitalocene. They argue that the industrial revolution was only possible through colonialization, slavery, and the general maleficence of the ruling elites. After creating a clear lead, the United Kingdom, the United States, France, and Germany established a global hegemony. And after the Second World War, it became too powerful to stop. Capitalist ideology ran rampant over the planet, making it difficult or impossible for nations with depleted resources, for example across the African and South American continents, to catch up.

This is a compelling narrative. But is there more to this story? The Golden Age of Capitalism ran from 1945 to the oil shock in 1973, when oil producers drove up prices four-fold. But it was not only capitalism that shone in this time. Several things happened. The Second World War destroyed most major economies except that of the United States. The United States knew it must rebuild other economies, otherwise no one would buy its stuff and this would stop growth. At the same time, the war had created enormous production capacity and new technologies, such as jet engines, radar, electronics, and computers, which could all be repurposed for civilian use. The United States and the United Kingdom used their new global dominance and the weakness of other nations to create new international institutions, such as the United Nations. The aim of the United Nations was predominantly to promote peace and settle conflicts, but also to promote growth, trade, and economic development. The newly established World Bank would lend money to countries to aid development, the International Monetary Fund would ensure financial stability in nations, and the precursor to the World Trade Organisation would promote international trade. In addition, European countries set in place the foundations for the European Union and abandoned colonies, although they left a trail of economic, political, and emotional destruction in their wake. All this helped support an unusual political stability across the globe – the Cold War notwithstanding.

But all this is still not enough to explain the Great Acceleration. Countries led by the United States and the United Kingdom reformed their economic policies. The 1920s saw the beginnings of consumerism at a significant scale, in the United States at least, with the application of behavioural psychology to weaponize marketing and advertising to create a new breed of consumers. But the Great Depression in the 1930s, sparked by a stock market collapse in 1929, spooked consumers, halted spending, and exposed the limitations of government policies to take a "hands off" approach to economic affairs. It took a second war to bring back more positive views of government interventions in the markets.

Advances in marketing and advertising reached maturity in the 1950s. At the same time, left-wing parties came to power in the United Kingdom and elsewhere in Europe and experimented with national healthcare and education systems on unprecedented scales. This improved health and social mobility, increasing longevity, and created the foundation for innovation with free and accessible university education. The capitalist machinery drove enormous wealth, but high taxes funnelled this wealth into supporting a rising middle class – a consumer class – and national and global infrastructure, such as motorways, universities, and hospitals, leading to a reinforcing feedback loop: as more people came out of poverty, economies grew, thereby allowing investment to bring even more people out. The Great Acceleration was driven by a mixed economy adopting features of capitalism and socialism.

You are here

The economic growth miracle has created a dominant narrative, believed and embraced by economists and political leaders, that economic growth at all costs is essential for the well-being, prosperity, and stability of nations. But material growth and energy use based on fossil fuels, which emit greenhouse gases to the atmosphere, cannot continue indefinitely. As the economist Kenneth Boulding told Congress in 1973, "Anyone who believes exponential growth can go on forever in a finite world is either a madman or an economist." Moreover, research now unequivocally shows that economic growth in wealthy economies is no longer linked to well-being: rich countries are getting richer, but people are not benefiting. However, the solution cannot be to end economic growth. With billions of people still living in poverty and misery, the world needs economic development. Our Earthly economic system is a phenomenally powerful tool – one of the most powerful in the solar system, it seems. We must harness this power to restabilize Earth and at the same time end poverty.

One thing that did not grow during the Holocene was our brain. In fact, since the start of the Holocene and the arrival of the agricultural revolution, our brain size actually shrank by 10 to 17 per cent, probably due to poor nutrition. It took until the end of the Holocene, and the start of the Anthropocene, for brain size to bounce back to its former glory, as human health, and particularly childhood nutrition, improved.

The Holocene began with one revolution, agriculture, and ended with two, scientific and industrial. In the blink of an eye, one species rose to such complete and utter dominance of the biosphere that we, first accidentally and now capriciously and maliciously, interfered with the internal workings of our planet's life support systems. We should proceed with caution.

ACT

II

THREE SCIENTIFIC INSIGHTS HAVE CHANGED HOW WE VIEW EARTH

Tipping points are so dangerous because if you pass them, the climate is out of humanity's control: if an ice sheet disintegrates and starts to slide into the ocean there's nothing we can do about that.

JAMES HANSEN
FORMER NASA RESEARCHER, 2009

As a scientist, the past 20 years have been quite overwhelming. We are learning so much about how our planet works, and the more we learn, the more reason we have for concern. We, humanity, are now such a large force of change that we have destabilized Earth. This is mind boggling. To remind myself of our new predicament, I always carry a small blue marble in my pocket, symbolizing our home planet, as a reminder that we are no longer a small world on a big planet, but have now become a big world on a small planet. So small, I can even, symbolically, have it in my pocket, reminding me all the time of our responsibility for taking care of the entire system. **Johan**

The COVID-19 pandemic in 2020 gave everyone on Earth a crash course in exponential growth. The number of infections doubled, then doubled again and again – the main feature of this type of mathematical function. At first, it seemed as if the coronavirus was a storm in a teacup. As China took unprecedented emergency action, essentially stopping its formidable economy, other countries downplayed the seriousness, even as cases began to appear within their borders. Only those countries that had faced recent epidemic risks with SARS – South Korea and Taiwan – really grasped the profound threat of exponential growth of the virus.

For the past 70 years, exponential growth in just about everything has been the dominant trajectory on Earth. It is unmistakable, yet often glossed over by historians more interested in world wars, cold wars, and cyber wars. The scale of growth is incomprehensible to many. As US nuclear physicist Al Bartlett once remarked, "The greatest shortcoming of the human race is our inability to understand the exponential function."

When it comes to exponential growth and Earth, the French lily pond riddle gets to the heart of the issue. One day, a pond has a single lily pad. The next day it has two, then four the following day, and so on, doubling daily. By day 30, the pond is completely full of lilies. On which day was the pond half full? Of course, if you go with your gut instinct, the answer is day 15. With a little more thought, it is, in fact, day 29. When you are on an exponential ride, you often do not realize that you are reaching a limit until you bust right through it. Imagine the lily pond on day 29. Everything seems open, uncrowded, and comfortable. The next day: bam! Saturation. Bursting at the seams. Breaking the boundaries.

Rebooting our relationship with Earth

Since the industrial revolution, we have followed the same social and economic logic. James Watt's improvements to the coal-fired steam engine inadvertently sparked a ferociously strong self-amplifying feedback loop to exponentially overexploit Earth's resources. If Earth were the lily pond, Watt brought us to day two and we are now at day 29 or 30.

Currently, we live on a planet with dangerously unstable life support systems. If we push the systems too far and set off more feedback loops, we will cross harmful tipping points and irreversibly start a transition from one state to another. Let's add some emphasis here. Until very recently – probably as recently as the 1980s or early

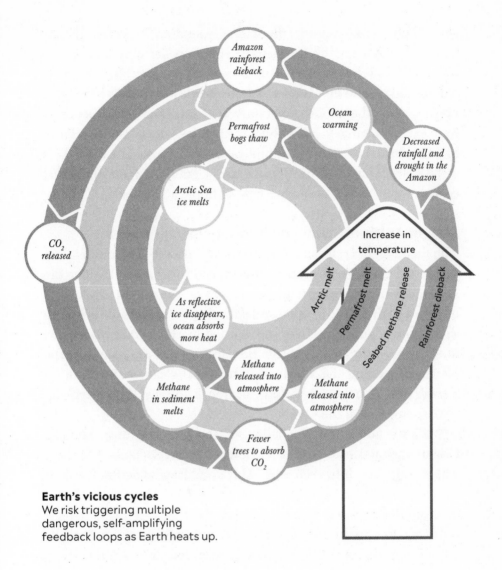

Earth's vicious cycles
We risk triggering multiple
dangerous, self-amplifying
feedback loops as Earth heats up.

1990s – the systems that regulate the state of the planet were still relatively stable. This was good for humanity; our modern development depends upon it. Destabilizing systems on Earth generates extreme events and puts us at risk of crossing dangerous tipping points.

This scenario is no longer a distant, potential threat. Right now, we have evidence that shows beyond any doubt that the Amazon rainforest and Greenland ice sheet are actively experiencing unprecedented change: deforestation and temperature rises are causing sleeping giants to wake up. Dear reader, whichever generation you belong to – Boomer, Generation X, Millennial, or Generation Z – Earth's life support system is destabilizing on our Earth watch. What happens next is dependent upon how we act.

How did we end up in this situation? While alarm bells have been ringing since the 1960s – certainly acid rain and the ozone hole were early warnings – it is only in the past two decades that science has established the profound implications of industrialization. This is the story of the three most important scientific insights of the 21st century, all of which have far-reaching implications for us all.

Insight 1: We are now in a new geological epoch

The first insight is the shocking upward trajectory of humanity on Earth: the Great Acceleration, introduced in Chapter 4. Since the 1950s, this has put exponential pressure on the planet.[1] Indeed, it has been so dramatic in scale and pace that there is ample evidence that we have exited the Holocene. We now live in the Anthropocene, in which we humans constitute the largest force of change on the planet, exceeding the variability and shocks caused by shifts in Earth's orbit, earthquakes, and volcanic eruptions. We are truly in the driving seat of the planet.

We have always faced challenges, though. The unsustainable irrigation practices of the first Mesopotamian societies left them with too much salt in their soils. Maya and Inca societies overused their resources – farmers could not feed growing populations, and deforestation and soil erosion reduced the resilience of the land – thus contributing to their collapse. There is evidence, too, that the fall of the Roman Empire was accelerated in part due to a combination of infectious disease, poor waste management, and heavy metal contamination in rapidly growing urban centres. Despite these islands of unsustainability and catastrophe, humans did not rock the whole planet. These were disasters, sure, but they were isolated occurrences. We, as humans, were gradually increasing

---

1 In climate science, the most iconic graph is called the "hockey stick", which shows just how sharply, and exponentially, the global average temperature has changed in recent decades. Earth system scientists do not have just one hockey stick; we have a whole team – from greenhouse gases to biodiversity loss and use of water, we see exponential trends wherever we look (see plate B2-3).

our footprint on Earth, but we did not yet have the numbers to do any global damage. We were still very clearly a small world on a big planet. In the mid-1950s, however, everything exploded. This was the take-off point for the exponential journey that has led to all the disastrous, negative impacts, from global warming to loss of nature.

The Great Acceleration graphs (see plate B2–3) show this exponential rise of pressures we are putting on the planet. We are out of control, and Earth's life support systems are lurching out of whack. This is what an increasingly turbulent Earth system looks like. Pick anything in the natural world that directly impacts your well-being. From natural resources, such as fresh water, soil, nutrients, and metals, to ecosystems like forests and grasslands, to the ozone layer to numbers of pollinators or fish in the ocean, you will find the same pattern: an unprecedented exponential rise pushing us ever closer to dangerous tipping points. Indeed, from the deepest ocean trenches to the edge of space, it is difficult to find anywhere on Earth untouched by humans. And the drama is that science shows the exponential pressures are now hitting the hard-wired, biophysical ceiling of what the planet can cope with.

In 2000, a scientific meeting was convened by the International Geosphere-Biosphere Programme (IGBP) in Mexico. It brought together the world's leading experts from a range of scientific disciplines, including geology, ecology, climate, agriculture, atmospheric physics and chemistry, oceanography, and many, many more. During the meeting, scientists from around the world presented up-to-date research on the state of the planet in the late Holocene. Listening intently sat the Nobel prize-winning chemist Paul Crutzen, who was a member of the IGBP committee at the time. Also in the room was Will Steffen, the director of IGBP. Steffen recalls that he kept glancing at Crutzen during one particular session because Crutzen seemed increasingly agitated.

At some point, Crutzen snapped. He stood up during a presentation and said, "Stop."

"Stop talking about the Holocene," he continued.

"We are not in the Holocene." Crutzen realized this needed some further explanation, but was at a loss.

"We are in the ... " he started.

"We are in the ... " he grasped for the word.

"We are in the ... Anthropocene."

People in the room recognized immediately that something important had just happened. Steffen recalls that the following coffee break was alive with discussion of this new concept.

Since 2000, the Anthropocene idea has been widely adopted across many academic disciplines. While it has yet to be formalized by stratigraphers, the Anthropocene is now very well scientifically established. Indeed, the International Commission on Stratigraphy has set up a special group to assess the case for this new geological epoch. This group of geologists, Earth system researchers, and other scientists has firmly concluded that there is an abundance of evidence for the Anthropocene, which means that, millions of years from now, future geologists will be able to piece together the story of this profound disruption.

One big debate, though, is the question: when did the Anthropocene begin? Initially, Crutzen proposed the start of the industrial revolution, but, as we saw in the previous chapter, the hockey stick graphs tell a different story. The rupture with the past, with stable Holocene conditions, really began around the 1950s.

Insight 2: The Holocene was astoundingly stable for 10,000 years

The second insight is what comes out of all the scientific evidence described in Act 1, namely that we, and our modern world, depend on the stability of the Holocene interglacial state of the planet for our own future.

In a foreboding twist of fate, we are gathering mounting evidence from ice cores, tree rings, and rocks that the Holocene was a uniquely stable state of the planet. Not only that, it is the only state of the planet that we are sure can support the modern world as we know it. We have lived outside the Holocene before, as hunter gatherers, in a deep ice age. But back then we were at most a few million people. There is absolutely no scientific evidence that Earth can support 10 billion people, with a minimum level of adequate quality of life, within a rapidly destabilizing life support system.

This is the drama. Just as we learn that we depend on the stable conditions of the Holocene, we find overwhelming evidence that we have left the Holocene and have entered an altogether more fragile and unpredictable era. Or looking at it another way, for two centuries we have been playing Jenga with Earth. We have been pulling out the blocks: the ozone layer, the ocean, the forests, the ice sheets. Now the tower is wobbling. Should we keep pulling out the blocks of wood, or should we start putting some back to stop the whole structure crumbling?

Insight 3: Earth's tipping points are too close for comfort

The third insight is that Earth has a remarkable resilience. It has taken our punches without hitting back. Go beyond certain thresholds, or tipping points, though, and beware, natural systems may go from being our best friend to becoming a deadly foe: difficult if not impossible to contain or control.

We have spent the past 70 years punching the planet. How has Earth responded? Well, Earth has behaved like Rocky Balboa (from the boxing films). Rocky rolls with the punches, gets knocked down, but staggers to his feet. Each time, his legs become more wobbly, his vision more blurred. He, like Earth, is very resilient: he absorbs the shocks without tipping over into a new state, although warning signs are there if you know where to look. By the ninth round, he is in such bad shape that a light tap pushes him over the edge and he crashes to the canvas. Rocky's brain has crossed a threshold. It has changed state, from conscious to unconscious.

Earth has a remarkable biological and physical resilience to deal with shocks and stresses without crossing a tipping point to a new state. But as we found out in Chapter 2, Earth has a hair trigger.

What exactly is a tipping point, though? In this context, a tipping point is the point at which a complex system like a brain or a rainforest crosses a threshold from a relatively stable state to collapse into new state. This can be caused by very small changes, like the Arab proverb, "The straw that broke the camel's back." Once a tipping point has been crossed, it is very difficult to pull back. Imagine a boulder at the top of a hill. One nudge might be enough to start it tumbling down, but it takes a lot more force to arrest its descent. If an ice sheet starts collapsing, reining in greenhouse gases will probably not save it (although it might slow down the inevitable).

In 2008, Tim Lenton, now at Exeter University, led a team to pinpoint the most important climate tipping points. They identified about 15 – from permafrost in Siberia to the ice sheets of Greenland and Antarctica, from ocean circulations to El Niño in the Pacific (which affects global climate) and the Amazon rainforest. Since their paper was published, more tipping points have been identified, such as the jet stream – a band of warm air above our heads here in Europe, North America, and Asia. In addition, Garry Peterson and Juan Rocha, at the Stockholm Resilience Centre, have created a huge map of tipping points. They have identified more than 300 examples of tipping points worldwide, from ice sheets to dead zones, from rainforests to permafrost. Tipping points are everywhere. But we are particularly concerned about the behaviour of just a handful of major systems that have looming tipping points.

What happens if we cross a tipping point in the climate system? Well, for a start we can forget about adapting incrementally to climate change. The rising sea levels will accelerate and become less predictable. Nature will start belching carbon into the atmosphere, as soils in the Arctic circle thaw, as fires burn in the Arctic, and as the Amazon swings from carbon store to carbon emitter. It will affect this generation and dozens of future generations until the planet finds a new equilibrium.

How high is the risk of crossing a climate tipping point? A decade ago, scientific assessments estimated that the risks become very serious when global temperatures exceed 4°C (7.2°F) global warming. Over the past 20 years, as we learn more of how our planet works, this figure has been creeping down. More recently, high risks are estimated at global warming above 2°C (3.6°F). Right now, we are at 1.1°C (2°F) and rising fast. In the past year or two, we have received worrying signs that tipping points are much, much closer than we once thought. This field of unexploded mines is not in the distance far down the road. We are standing right in it and, as we will discuss in Chapter 9, we now need to tread extremely carefully.

The sum is greater than the parts
These three scientific insights have fallen into place only in the past five to 10 years, thanks to the extraordinary advancements in science over the past four decades. They involve progress in

different disciplines, from climate science and economics and ecology to glaciology, oceanography, and anthropology. But they also include the bringing together of all the sciences to reveal how our planet works: the sum is greater than the parts. As James Lovelock and Lynn Margulis postulated in the 1970s with the Gaia theory, the Earth is truly a complex self-regulating system, where all living and non-living components of Earth are connected and, through interactions, collectively determine the final state of our planet.

Even with this knowledge, we continue to exert ever greater pressure on our planet's life support systems without a care in the world. We can no longer ignore the fact that this pressure may trigger disastrous, unstoppable, and abrupt changes if we continue on our current path. The only conclusion is that we must step very carefully into the future. Instead, we are behaving like a herd of stampeding elephants.

Perhaps there is a final, bonus insight in this chapter, though: exponential growth cannot go on forever. Eventually, the steep slope bends and flattens. We have come to the end of the road of our conventional economic paradigm of exploiting nature for free and discarding our waste. Our future now depends on our ability to transition to a new logic of economic development, prosperity, and equity within a safe operating space on Earth. We have reached the edge of the lily pond.

PLANETARY BOUNDARIES

It is time for us to step up
And respect the boundaries
Of just how far we can push this planet
Become stewards of our collective futures
And recognise just how much our
livelihoods depend on it
We live in a globalised community: a big
world on a small planet
Where every flutter of butterfly wing can
either serve to strengthen the hurricane
Or fuel the winds of change.

ALINA SIEGFRIED
STORYTELLER, 2018

It is 2009, six months before the catastrophic Copenhagen climate summit and a few weeks after I arrived in Sweden. Kevin Noone came to my office at the International Geosphere-Biosphere Programme. Kevin, a former director of the programme, is a climate scientist. He mentioned that he had a paper coming out in *Nature*[2] shortly, with Johan and a bunch of other researchers, about something called "planetary boundaries". "That's interesting," I said. But I didn't mean it. Scientists had been talking in quite abstract ways for several years about the Earth system and its myriad complexities.

---

2 If you discover life on Mars, or the source of dark matter, you publish your findings in one of two scientific journals: *Nature* or *Science*.

Kevin said that they had identified nine boundaries that keep Earth stable. I mulled this over. Nine. That is quite interesting. Not 100. Not 1,000. I was genuinely intrigued now and asked, "Do you quantify them?" I thought I already knew the answer. Scientists like abstract models and frameworks. They rarely put numbers to them, especially not in Earth system science. It is just too difficult. Too many unknowns. This ensures most ideas stay in academia, rarely troubling the world outside.

But Kevin said something that completely floored me: "Yes."

He was humble about it. He emphasized that a lot more work needed to be done. But the boundaries were quantified. Now Kevin really had me hooked. This is a game changer for the planet.

The planetary boundaries framework is a remarkable scientific achievement. Eleven years after publication, there is still no better way to think about a stable planet. For the first time, the world had a priority list for the Earth system. **Owen**

It is only in the past 10 years or so that science has armed us with three important insights that show that humans now constitute a force of change of geological proportions on the planet. We are at risk of tipping the entire planet out of balance and bringing an end to the stable Holocene state that has supported us for more than 10,000 years. It is difficult to overemphasize the drama. There is no doubt in our minds that this fundamentally changes how we go about our daily lives. But can we provide scientific guidelines to safely navigate the future?

We are already hitting the ceiling of hard-wired processes that regulate the stability of Earth. If tipping points are real, then there are points of no return. This means that we need to do everything we can to keep the planet in a state in which we have two ice caps and stable sea levels, forests and wetlands that store carbon, a climate that reliably supports agriculture, and ocean currents that distribute heat predictably: a Holocene state. To do this, to create a planetary boundaries framework, we need to know the following.

First, what are the processes and systems on Earth that regulate the state of the planet? Once we have identified them, we need to think about quantifying them. We must somehow identify the danger zone, where we risk triggering uncontrollable, irreversible change. In short, we want to identify a safe operating space for humanity, within which we have a good chance of keeping Earth in a stable state. This is a tricky, fiendishly complex scientific challenge.

Second, what kind of planet do we need for our future? Biological, chemical, and physical processes and systems regulate Earth's state. Can we quantify the boundaries related to these processes to support 10 billion citizens? Here, we should emphasize that the scientific process of identifying planetary boundaries does not care about us humans. For the academic exercise, we care only about Earth and the search for the biophysical processes that determine what it will take to keep our planet in a livable state.

The planetary boundaries give us a safe operating space, within which we humans can thrive and prosper. They give us room to sort out challenges of equity and redistribution of wealth, pursue happiness and peace, improve health and safety. If we push beyond the planetary boundaries, then all this becomes far more difficult. We will be trying to cope with the trials and tribulations of everyday life while dealing with a dangerously destabilized planet. We will be fighting poverty and hunger, inequality and disease, while tackling ever worsening heatwaves, droughts, floods, and rising sea levels.

In 2007, we[3] invited researchers to Sweden to map all the systems and processes that regulate the state of the planet. We turned over every scientific stone we could find. After scanning the Earth system landscape for evidence, we found nine processes and systems and quantified[4] seven (see plate B4). These include climate, biodiversity, ozone, fresh water, and land. Alarmingly, we estimated that three boundaries have already been crossed. Earth is already in the danger zone. We will go into detail in a moment.

In 2009, we published the paper in *Nature* for scientific scrutiny. This is how science works best. You put out an idea and make the strongest case you can, based on the available evidence. Then, you subject it to the harshest critics – your peers – who have the breadth and depth of knowledge to find holes in your arguments and destroy

---

3 The "we" here refers to Johan. Owen moved to Sweden in 2009.
4 We used science to define the variables that regulate the state of that specific planetary boundary and put a number on them.

your case. The paper caused an intellectual firestorm. Research groups looked at each boundary and proposed improvements and iterations. "Where is soil?" some asked. "You missed plastics," argued others. In 2015, after six years of intense scrutiny, we reassessed the state of knowledge. Had we missed any fundamental process contributing to the state of our planet? Was there scientific evidence questioning any of our proposed nine boundaries?

We published the updated paper in the journal *Science*. We concluded that our first hunches were correct: the nine boundaries that we had identified in 2009 still remained in 2015. We did not find enough evidence to include soils or any of the other suggestions (plastics might fall under novel entities when that boundary is quantified). It is really reassuring to be able to say in 2020, more than 10 years after the first presentation of the planetary boundaries framework, that we have identified the right nine boundaries. We may still debate and improve on the numbers as science advances, but we can conclude, with a high degree of scientific certainty, that our approach works. And as long as we get it right on these nine processes and systems, we stand a good chance of getting it right for the planet, for us, and for future generations.

So, if it isn't soil and plastics, what are the nine boundaries?

The big three

There are three gigantic systems that operate at the planetary scale, which have known tipping points and regulate the state of the entire planet. They are (1) the climate system, (2) the ozone layer, and (3) the ocean.

1. The climate system

The climate system connects the ocean, land, ice sheets, atmosphere, and rich diversity of life. It controls how much water is locked as ice in glaciers and polar ice sheets, which in turn regulates whether sea levels fall or rise. Sometimes, these changes are extreme. During a deep ice age, the sea level can drop by 120 metres (395 feet), but if temperature climbs much further we lose Antarctica and Greenland and levels rise by 70 metres (230 feet). The climate system controls what we can grow and where we can grow it – and whether agriculture is possible.

We know today that the average temperature on Earth has fluctuated by only 1°C (1.8°F) for 10 millennia. This stability is how we characterize the Holocene state. The million-dollar question, though, is how high can we risk temperature rising? If we stray too far, Earth may cross a climate tipping point and start moving along a new unstoppable path away from a Holocene state. We could end up in a much, much warmer hothouse Earth state (see Chapter 7). We do not have the exact answer to this question. It is one of the grand scientific quests of today. However, we do know enough to say that we will cross some tipping points between 1 and 2°C (1.8 and 3.6°F) global warming. Tragically, it will be time to say goodbye to coral reefs and at least one ice sheet (in Antarctica).[5] Ancient evidence indicates that up to 2°C (3.6°F) Earth does not cross a climate tipping point. The risk level, though, is not zero.

We are, today, at 1.1°C (2°F), and we are starting to see the impacts. Given the record-breaking temperatures, phenomenal ice melt, coral reef death, and falling carbon sink in the Amazon in the past two decades, there is strong evidence to support placing a climate planetary boundary around 1.5°C (2.7°F), at a safe distance from the tipping point. To have a high chance of staying well below 1.5°C (2.7°C), given all the uncertainties, we recommended that the world should keep carbon dioxide level in the atmosphere to about 350 parts per million (ppm). This defines the climate boundary. Today, we are at 415 ppm. We are already in the danger zone. Beyond 450 ppm the world enters the high risk zone. We are on very thin ice.

In 2015, nations met in Paris and agreed to keep global warming well below 2°C (3.6°F) and to aim for 1.5°C (2.7°F). There is consensus in the scientific community that this is a reasonably safe boundary.[6]

2. The ozone layer
Our protective shield, the ozone layer, safeguards life on Earth's surface by absorbing dangerous levels of ultraviolet radiation from

---

5 Some tipping points lead to inevitable change, such as sea level rise or ecological collapse, which may not directly impact carbon emissions. Others turn carbon stores into emitters, thus risking a runaway climate impact.
6 In 2018, the Intergovernmental Panel on Climate Change concluded that 1.5°C (2.7°F) is considerably safer for humanity than 2°C (3.6°F).

the Sun. Without it, this radiation damages DNA in plants, animals, and humans, and also causes skin cancer.

Few know how perilously close humanity came to destroying the ozone layer in the 1980s. The story began back in the 1930s when chemists invented a new class of chemicals, CFCs, containing chlorine, to improve how refrigerators and air conditioning work. Ironically, CFCs were thought to be safer than the volatile alternatives used at the time. Unfortunately, no one guessed that these chemicals would drift high into the atmosphere, where they would annihilate the ozone layer. In the 1970s, scientists connected CFCs to ozone damage, but lacked evidence that this was a big problem. In 1983, scientists working for the British Antarctic Survey raised the alarm: a massive hole over the entire continent had formed abruptly. We were in deep shit. We would be in much deeper shit had bromine been used instead of chlorine. The elements are interchangeable, but – atom for atom – bromine is 45 times more ferocious than chlorine when it comes to destroying ozone.

The ozone layer is measured in Dobson Units (DU), which relate to its thickness in millimetres. This planetary boundary is set at a minimum thickness of 275 DU. Earth is currently within the ozone boundary, which was not the case in the 1980s. After decades out in the high-risk zone, we were able to return to a safe operating space. Political leaders acted to save Earth and created the Montreal Protocol on Substances that Deplete the Ozone Layer in 1987.

The ozone hole still exists today, but it has stabilized. Use of CFCs fell 57 per cent in a decade after the protocol came into force in 1989.[7] The ozone layer is expected to make a full recovery by around 2060. This story is an example of the speed, scale, and surprise we face in the Anthropocene. We undoubtedly had a lucky escape.

3. The ocean
We live on a blue planet. The ocean covers 70 per cent of the surface. It is so vast and seemingly infinite that it is easy to think it can cope with anything we throw at it. The ocean – there really is just a single connected ocean – is, in large part, the engine room of

---

[7] In recent years, scientists noticed CFCs on the rise again and traced the violation back to unscrupulous factories in China.

the planet. It regulates heat exchange between the atmosphere and the surface, hosts a huge diversity of life, and regulates the flow of nutrients and the water cycle. A stable, well-functioning ocean is a fundamental prerequisite for a stable, well-functioning planet. It stores 93 per cent of the heat caused by our burning of fossil fuels. The observed 1.1°C (2°F) global warming is only a small fraction of the energy imbalance caused by us humans. If all the heat absorbed in the ocean from our emissions was suddenly released into the atmosphere, the global temperature would momentarily rise some 27°C (48.6°F). There is no science suggesting this will happen, but it shows the power of the ocean.

We want the ocean to keep doing what it has always been doing. The heat in the ocean is captured by the climate planetary boundary, so the ocean boundary is based on ocean acidification. As a result of carbon dioxide emissions from burning fossil fuels, the acidity of the ocean has changed by a mind-boggling 26 per cent since the start of the industrial revolution. The only thing that comes close to this is the Paleocene-Eocene Thermal Maximum shock 55 million years ago – a major extinction event, particularly in the ocean – but that happened over a much longer period. The rate of ocean acidification we are seeing today is significantly more rapid, and if this continues for long the results are certain to be catastrophic. Ocean acidification makes it difficult for ocean life to grow hard shells or calcium carbonate skeletons, such as corals, oysters, and the vast mass of calcifying phytoplankton floating on the ocean surface. Oyster farmers are already affected by this, and corals are dying, as heat, ocean acidification, and pollution continue to increase. In 2020, as another heatwave wrecked the Great Barrier Reef, scientists warned that Earth may have crossed a tipping point with regard to the annual bleaching of corals. At the moment, Earth remains within the ocean acidification boundary. Realistically, the only way to stop ocean acidification, though, is to halt fossil fuel emissions.

The four biosphere boundaries

Tightly connected to the "big three" are the four biosphere boundaries. There is no strong scientific evidence that exceeding these boundaries will lead to planetary-scale tipping points.

However, the biosphere boundaries are still critically important. They moderate Earth's life support systems by amplifying or dampening the boundaries of the big three. The biosphere boundaries regulate the resilience of our planet.

The four biosphere boundaries are (1) all living species on Earth and how they connect – what we call biosphere integrity, (2) the critical biomes, or the big natural ecosystems on Earth – from rainforests to tundra with savannas, wetlands, and boreal forests in between, (3) the global water cycle, and (4) the global flows of nitrogen and phosphorus, the key biogeochemical cycles. Biodiversity, land, fresh water, and nutrients operate at the smaller scale of ecosystems and watersheds but aggregate up to the planetary scale. They function as the ultimate insurance against Earth leaving its safe operating space.

1. Biodiversity

Living nature – that is, all microbes, plants, trees, and animals on land and in the ocean – supports Earth's stable state. As we saw in Chapter 1, it is the living planet that enables the physical systems on Earth to remain in Holocene equilibrium. The trees moderate greenhouse gases and maintain rainfall patterns: much of the rain falling in the middle of the Amazon comes from water that has evaporated from leaves and trees farther east – the forest recycles water. If we lose them, Earth could tip away from the Holocene state.

Biodiversity plays two fundamental roles for human life on Earth. First, its genetic diversity provides Earth with its long-term capacity to resist change. It is life itself that moderates extremes. Earth is in a Goldilocks state thanks to the diversity of life. Second, biodiversity – the composition of living organisms – determines the final type of ecosystem and all its functions. It is the diversity of plants, animals, and microbes that gives us a rainforest system, and it is the diversity that allows a rainforest to keep doing what rainforests do. A large variety of different species all providing the same function, such as pollination or water purification, is what gives nature its resilience and its capacity to deal with shocks.

Earth is now beyond the boundary for biodiversity. This is more than just the shocking numbers of species going extinct; it also relates to the integrity of ecosystems – how species interact. We are shredding the fabric that gives the Earth system its deep resilience.

2. Land

How much forest do we need? How much wetland? We need to ask these profound questions, but we do not have all the answers. However, we can say for certain that we have already altered half of the land surface of Earth and we manage about 75 per cent.

The US ecologist E. O. Wilson advocates for a "half-Earth" conservation strategy, meaning we need to set aside half the planet for biodiversity and natural ecosystems. Wilson is right. The balance of different ecosystems, including forests, meadows, wetlands, peatlands, and tundra, moderates the Earth system. How far can we push them? It turns out that a critically important factor is biomass. The state of the planet boils down to the extent of our largest forest systems: namely, our three remaining rainforests, in Amazonia, the Congo, and Indonesia, and the temperate and boreal forests in the northern hemisphere.

The land planetary boundary is defined as the percentage of forested land compared with original forest cover. As a simple rule, we should keep as much as 75 per cent original forest cover. Currently, the global area of forested land is 62 per cent of original forest cover. We have overstepped the boundary. More generally, we must at all costs protect the remaining 50 per cent of Earth's surface and regenerate the areas we have previously destroyed.

3. Fresh water

Water is the bloodstream of the biosphere. Everywhere that liquid water has been found, we have also found life.[8] Water determines whether life thrives or dies. It distributes nutrients and other chemicals essential for life, and it is fundamental to photosynthesis, where water is split into oxygen, thus enabling all biomass growth.

---

8 This fact alone makes the discovery of liquid water on any planet or moon in our solar system or beyond very intriguing. We have evidence that water flows intermittently on present-day Mars, and liquid water may exist below the icy crust of one of Saturn's moons, Enceladus, and one of Jupiter's moons, Europa. Space agencies should establish missions to these places immediately!

If all fresh water on Earth just disappeared, our planet would die. If you were to slowly add fresh water back, drop by drop, at some point life would return on Earth. Eventually, we would cross a tipping point and a new biosphere would re-establish. How much fresh water needs to flow for Earth to kick-start its biological life-giving processes? We do not even need to be this extreme. Instead, we only need to ask ourselves at what point does the ecosystem stop functioning in the same way?

Put differently, how much water can we take from watersheds, river basins, and ecosystems before they collapse? We need water for irrigation, industry, and domestic use. If we take too much, though, we may cross a tipping point and the ecosystems will not function well. The planetary boundary assessment estimates that we can consume up to 10 to 15 per cent of the total run-off on Earth. For water in rivers, we can use up to 50 per cent, although once we use more than 40 per cent of available water in river basins, we run into severe water stress. This defines the water boundary, and at a global scale we have not stepped over it, yet. But if you look at many river basins around the world, you will find a different story: locally, the fresh water boundary has already been crossed in many places.

4. Nutrients

In the past 50 years, something remarkable happened on Earth. Famine, a spectre stalking our civilization, all but disappeared. This was due in no small part to the invention of manufactured fertilizers based on nitrogen and phosphorus. Crops need water and sunlight, but they also need nutrients – nitrogen, phosphorus, and potassium – to thrive. Although the atmosphere is 78 per cent nitrogen, this cannot be used by most plants. Before the industrial revolution, farmers relied on naturally occurring "organic nitrogen", such as ammonia or nitrate, which is created by some plants and microorganisms and is widely available in manure. Now, industrially produced fertilizers support half of the global population – more than 3 billion people. But these fertilizers are very often not used efficiently. They are dumped on the land, causing massive pollution in soils, lakes, rivers, and coasts. This pushes the natural processes out of whack and flips ecosystems from being diverse and healthy to being dead zones.

The nitrogen and phosphorus cycles are two important parts of global biogeochemical cycles, and Earth is now far beyond the planetary boundary for both nitrogen and phosphorus use. Indeed, we are already experiencing the most significant change in the nitrogen cycle in perhaps 2.5 billion years.

We need nitrogen and phosphorus to feed the world, so it is inevitable that there will be high and rising pressures on the nutrients planetary boundary. Several recent research studies have shown, though, that it is possible to feed 9 to 10 billion people within planetary boundaries, including nitrogen and phosphorus. The challenge is to recognize two key things. First, we must acknowledge that farmers in rich and several emerging economies overuse fertilizers. They need to drastically reduce this usage, which will open up space for sharing with farmers in developing countries, who are using too little fertilizer and struggle to produce enough crops for food security. In short, the scientific definition of a safe boundary for nitrogen and phosphorus translates to an equitable boundary of sharing the remaining nitrogen and phosphorus budgets on Earth in a fair way. Second, the challenge is to transition production practices from the current linear systems – in which too much fertilizer is loaded into ecosystems, and allowed to leak into waterways and ecosystems downstream – towards tightly managed production, where precision application (only applying what the crop really needs) is combined with improved crop rotation (with plants that capture nitrogen and phosphorus in the soil). Most importantly, production must become circular. This means returning all surplus nitrogen and phosphorus back into the farm where it came from. This must include not only manure, but also waste from urban areas where the food is consumed.

What is so exciting is that these practices are well known, and have already been developed for both high-tech, satellite-based systems in wealthy economies and sustainable practices for small-holder farmers in developing countries.

The two aliens

Finally, we have identified two alien planetary boundaries. They did not exist in the Holocene, or indeed at any time in Earth's 4.5-billion-year history. They have been introduced by us and now interact in profound and unexpected ways with Earth's life support system. They are "novel entities" – a catch-all term for the thousands of chemicals we have developed and released – and aerosols, small particles in the atmosphere causing air pollution. What could go wrong?

1. Novel entities

Some substances were deliberately created by governments to cause harm and devastation. We are talking about chemical, biological, and nuclear weapons. In 1945, the United States carried out the first nuclear test, followed by the Soviet Union in 1949, the United Kingdom in 1952, France in 1960, and China in 1964. At the time, there was limited interest in the health impacts, let alone long-term environmental threats, even though each nuclear test deposited radioactive particles everywhere on Earth's surface. It turns out that this is quite useful for future geologists, who will find an unmistakable signature for a major rupture on Earth: it coincides with the start of the Great Acceleration and Earth's arrival in the Anthropocene. In 1963, the Nuclear Test Ban Treaty outlawed all nuclear weapons tests in the atmosphere, space, and underwater. Now, only weapons testing underground is permitted.

Some substances are created by companies to solve one problem but end up causing more, such as the ozone-eating CFCs mentioned earlier, or adding lead to petrol to improve engines only to find that it has terrible effects on people's health,[9] or DDT, the first modern synthetic insecticide. If the world had just a handful of artificially produced chemicals to deal with, then we might have a manageable situation. But there are more than 100,000 man-made substances in the environment. These range from nuclear waste, pesticides, and heavy metals to plastics, including microplastics and nanoparticles.

We have limited knowledge of the risks that accumulate when novel entities build up in the biosphere and interact. Evidence suggests, though, that the cocktail effect of accumulating novel

---

9 In the 1920s and 1930s, the research to develop CFCs and to add lead to petrol was led by the same person: US chemist Thomas Midgley. The environmental historian J. R. McNeill once said Midgley "had more impact on the atmosphere than any other single organism in Earth's history".

entities is likely to trigger unwanted and unstoppable change. What is the limit, and where is the safe operating space? We are still trying to figure that out.

Novel entities also include emerging risks such as artificial intelligence (AI). In this field, there is a famous thought experiment, called the paperclip maximizer, about the risk of a superintelligence adopting an arbitrary goal and how this might escalate. Imagine that a CEO needs a paperclip to fasten some papers together. He says to his advanced AI, "Hey Iris, make sure we don't run out of paperclips ever again." Iris diverts the entire business to this task, eventually identifying new revenue streams to devote to paperclip purchase. A new economy springs up solely to supply the AI's demand for paperclips. Eventually, the AI buys paperclip wholesale companies, mines, and foundries. Increasingly, more of the world's resources become devoted to paperclip supplies. And so on. While seemingly preposterous now, the point is that there is an infinite number of ways in which the goals of a superintelligence might deviate from those who programmed it, or those who are impacted by it. It would be helpful if AI is cognizant of the state of the planet, among other things.

Genetic risk also falls under novel entities. Take gene drives: deliberately releasing an altered gene into a wild population, for example to eradicate mosquitoes carrying malaria or the Zika virus. The temptation to do so is strong. Malaria kills almost half a million people every year, mainly in Africa, and the Zika virus can cause children to be born with birth defects. Obviously, there is the possibility that gene drives in the wild will go without a hitch, and humanity will be rid of these dreadful diseases. But there are a thousand reasons why you would want to think twice or even three times about such action.

Given the potential complications and with so many unknowns, the novel entities boundary has yet to be quantified.

2. Aerosols
A brown cloud stretches for thousands of kilometres over China and India. A dense fog of pollution hangs for weeks over Beijing, Delhi, and other Asian cities. Sometimes the pollution mingles with smoke from forest fires in Indonesia. Forest fires in California,

Australia, Canada, Scandinavia, and Russia throw up vast soot clouds that drift for weeks. This is all part of life in the Anthropocene.

Industrial processes, car exhausts, fires, and the burning of fossil fuels release small particles called aerosols into the atmosphere. The accumulation of aerosols leads to air pollution, which causes up to 9 million premature deaths every year. Aerosols also affect how the planet works. They influence cloud formation, allowing more clouds and rain to form near cities. They also affect weather patterns. Some aerosols absorb heat, while others create a layer of haze that prevents solar radiation from reaching the surface of Earth. Aerosols released from cooking and heating with biofuels and diesel transportation in India may be making the Indian monsoon irregular and bringing less rain. India needs rain for the crops to feed its 1.3 billion people.

The aerosols planetary boundary is about keeping regional weather systems working, but it is also deeply linked to climate. There is much uncertainty and we have yet to calculate a precise global boundary. However, we have come quite a long way for one of the key hot spots on Earth – the South Asian monsoon – where we have been able to propose a boundary related to the maximum allowed haziness in the lower atmosphere.

We are in deep danger

We have transgressed four of the nine planetary boundaries: climate, biodiversity, land, and our use of nutrients. This should alarm everyone. We could now cross a tipping point in any one of these systems. For biodiversity loss and nutrient overloading, the risk of triggering unpredictable and irreversible changes is higher. We are at red alert. Meanwhile, climate change and land system change have reached the danger zone. We are at amber alert.

We can see this with our own eyes. At 1.1°C (2°F) global warming, the frequency and amplitude of extreme weather is increasing. We have reached mass extinction rates of species. Indeed, around 1 million of an estimated 8 million species are now threatened with extinction. Devastatingly, since 1970, we have reduced animal populations by 60 per cent. Fires and drying of the Amazon rainforest, bark beetle outbreaks in Canadian boreal forests, and drought-induced fires in European temperate forests are just a few indications of turbulence in natural land systems. We should be tiptoeing into the future. Instead, we are blindly charging forward.

Our planet operates as a tightly connected system. It is similar to the human body with its interlinked organs. Your overall health relies not only on the functioning of your heart, lungs, liver, kidneys, and nervous system, but also on how they interact and communicate with each other. And importantly, just like humans, Earth seems to live by the motto "All for one, one for all". If we lose one organ, our entire system can shut down because that organ influences the state of all the other functions. So it is for Earth.

We do not know the exact position of each planetary boundary. Not because it does not exist, but rather because of the complexity of how Earth, with all its organs, operates. We are learning all the time. As knowledge improves, we become more nervous. Improved supercomputer capacities show that the climate is even more sensitive than we thought a decade ago. We cannot rule out that if carbon dioxide in the atmosphere doubles, Earth's temperature will rise to a catastrophic 5°C (9°F) global warming or higher. It is hard to imagine civilization coping with such a shocking transition. The upshot of this revelation is that the 1.5°C (2.7°F) target has become even more important to meet, but more difficult to achieve.

Should all planetary boundaries be treated equally? No, there is a hierarchy. Climate and biodiversity boundaries are core boundaries: on their own, they can push Earth into a new state. Not only that, these two boundaries depend on many other boundaries. The land, water, and biogeochemical flows determine the pattern of species. The deep currents in the ocean and the ice sheets, together with carbon, methane, and other gases in the biosphere, determine the final state of the climate. Transgressing one or more of the non-core boundaries may severely affect human well-being and trigger the transgression of a core boundary, but they cannot, as far as we know, push the Earth system into a new state on their own. Climate and biodiversity can.[10]

---

10 Increasingly, evidence shows that ocean acidification has been the prime cause of some of Earth's most severe mass extinctions.

HOTHOUSE EARTH

What is the difference between a 2°C world
and a 4°C world? Human civilization.

HANS JOACHIM SCHELLNHUBER
FOUNDING DIRECTOR OF THE POTSDAM INSTITUTE
FOR CLIMATE IMPACT RESEARCH

Let's go back to the summer of 2018. If you live in the northern
hemisphere, you may recall something unusual. The entire
hemisphere sweltered under an uncanny and unprecedented
heatwave. No region was immune. In Sweden, this was a never-
ending summer. It arrived in early May and stayed until late
September. Wells and groundwater dried out, causing the first
real water crisis. The forests dried out, too, and fires raged. This
pattern repeated across the northern hemisphere. In the middle
of this heatwave, the US academic journal *Proceedings of the National
Academy of Sciences* published our "Hothouse Earth" research paper
by Johan and colleagues. The media attention took us by surprise.

Some thought that the timing of publication had been designed
to coincide with the heatwave. If only we were that clever, or that
science worked to media deadlines.[11] In fact, the day the paper was
published, the lead author, Will Steffen, was returning from the
Galapagos Islands and was completely inaccessible to the media.

At some point on 6 August 2018, we crossed a tipping point.
Luckily, this was not a tipping point in Earth's life support system;

---

11 In recent years, science has advanced so far that researchers have been able to
produce research attributing human-caused climate change to extreme events,
such as hurricanes, within days of the catastrophe.

it was a tipping point in the media. The "Hothouse Earth" paper had been published earlier that day, and there was a flurry of interest from science journalists. They wanted to know whether humanity might inadvertently set off a chain reaction in the climate system, thereby making it all but impossible to restrain climate change at tolerable levels for civilization.

Throughout the day, the "Hothouse Earth" story kept getting bigger and bigger. Suddenly, we were not only getting attention from the science desks, but also from the main news desks. Requests flooded in. Once major media such as the BBC and CNN became involved, the story crossed a tipping point, which set off a cascade of other media interest around the globe. For a frenetic 24-hour period, the phones did not stop ringing. The paper struck a nerve and went on to become the biggest climate science story of the year. The word "hothouse" was awarded "word of the year" in Germany.[12]

We first introduced the hothouse Earth state in Chapter 1. It was a remarkably stable state, lasting millions of years. During this time, Earth was 4°C (7.2°F) or more warmer than it is today, dinosaurs roamed, there was no ice at the poles, and the sea levels were 70 metres (230 feet) higher than they are now. A hothouse world is very different from our world today. And, as Hans Joachim Schellnhuber[13] reminds us, the hothouse state is the difference between a stable civilization and collapse.

The "Hothouse Earth" paper begins with the observation that when fossil fuels stop flowing into the atmosphere, the temperature may not stabilize quite as we would wish. At the moment, the best estimate is that if nations keep to their commitments on fossil fuels, temperature will stabilize at 3°C (5.4°F) global warming. But we should be wary of this. We could reach a point where the temperature keeps spiralling even after emissions cease. Could we accidentally cross a tipping point that would send us irrecoverably back to the hothouse? The simple answer is yes. In this chapter, we will talk about how this scenario could unfold.

---

12 The paper was actually more mundanely titled "Trajectories of the Earth system in the Anthropocene", but it took on a life of its own in the media.
13 It is no exaggeration to say that Schellnhuber is an icon in Germany and one of the leading thinkers in climate research since the 1980s.

In Chapter 2, we discussed how the small changes, Earth's hair trigger, can send Earth spinning. Very slight alterations in Earth's orbit are enough to nudge Earth across a tipping point and send us from warm interglacial to ice age and back again. We have bobbed back and forth like this for the past 3 million years.

The transition from ice age to warm interglacial starts when more heat from the Sun reaches places such as Scandinavia and northern Canada in summer. This alone might warm Earth a few degrees, but not enough to lift it out of an ice age. It is enough, though, to kick-start the biological engine on Earth. The slightly warmer temperatures release carbon dioxide from the ocean, and permafrost thaws, thus emitting carbon, too. These carbon dioxide emissions drive the temperatures even higher. As ice recedes, plants colonize the north. Their darker colour absorbs more heat than white snow, thereby trapping more heat. This, too, causes even more warming and even more melting ice. Nature wakes up, changing the flow of carbon around Earth. This drives a self-reinforcing warming cycle, until nature eventually runs out of steam and temperatures settle about 5°C (9°F) warmer than the deep freeze.[14] Goodbye ice age.

We know for certain that Earth has responded quite profoundly time after time to small nudges, so why would it not happen again? But the Anthropocene is less of a nudge and more of a shove, in the wrong direction.[15] Although our ocean, land, ice sheets, and atmosphere respond to changes in heat from the Sun, this alone does not determine the final state of the planet. The biosphere – life – also gets in on the act in a surprising and often beautifully rich and complex way, speeding up change once thresholds are crossed, dampening shocks and stress when biophysically able. This ability to keep Earth in a stable state is what we call Earth resilience.

---

14 Bizarrely, climate sceptics use an Orwellian reasoning to confuse people over ice ages. They claim that because temperature rises before carbon dioxide levels increase, then carbon dioxide is unimportant. If they dived deeper into academic literature, they might be blown away by the wondrous way our biosphere works.

15 The difference between the last ice age and the Holocene is 100 ppm of carbon dioxide. We have already added an additional 135 ppm to the atmosphere in two centuries.

Nature can push and pull Earth's systems in different ways. At times, nature helps Earth roll with the punches by restraining a big shove and dampening it down. A pulse of carbon dioxide can be absorbed by trees and ocean plankton, for example. Without the ocean and soils and forests, we would have double the carbon dioxide in the atmosphere. This would be catastrophic. The biosphere is our friend; at least, it has been up to now.

However, the biosphere can amplify as well as dampen. This is our nightmare. If one tipping point is crossed, how does this affect all the other tipping points? This is a critical question for our future and the central focus of our "Hothouse Earth" paper.

The domino effect

The first comprehensive analysis of Earth's tipping points only appeared in 2008. Since then, the race has been on to quantify where the tipping points lie. At what degree of global warming will they tip? How much biodiversity loss is too much? Much less attention has been paid to how one tipping point might link to another. Even most climate models in use at the moment do not include all known tipping points. So, it is hardly surprising that we know precious little about how they interact. But we are beginning to get a handle on it. We call it the domino effect.

Let's take a tour of the globe and talk through a couple of "what if" scenarios.

The Arctic

We'll start high in the Arctic Circle. What if sea ice continues shrinking and thinning each summer? More and more dark water will be exposed beneath. The dark water absorbs more heat than the ice, thinning the ice even more, so more ice melts the following year. As the Arctic warms, vast fields of permafrost in northern Canada and Russia start to thaw, releasing potent greenhouse gases. At the same time, the warmer temperatures dry out the Canadian, Alaskan, and Russian forests and peatlands. This makes them prone to fires. Imagine, peatland fires on the few areas of Greenland without ice. Well, we do not have to imagine anymore; they are happening. But that is not all we need to worry about in Greenland. As glaciers collapse, large volumes of fresh

Cascading tipping points
Some parts of the Earth system are
prone to tipping points. Crossing one
tipping point can push Earth across
others, like cascading dominoes.

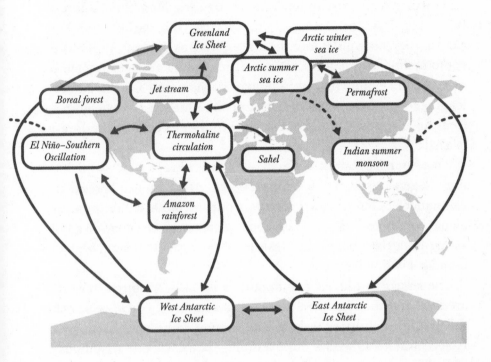

water flow into the North Atlantic, disrupting the ocean's circulation. This changes how heat moves around the planet and causes warm water to build up around Antarctica, melting glaciers from below and pushing them past a point of no return. We risk a cascade like this occurring when temperatures reach 1 to 3°C (1.8 to 5.4°F) global warming.

Remember, we are now at 1.1°C (2°F). We have already stumbled into the danger zone.

The Amazon

The changes in the Arctic and ocean circulation may also affect rainforests such as the Amazon, drying them out. Atlantic ocean currents greatly influence rainfall in the Amazon and the Sahel region of Africa. What if the Amazon turns from lush tropical growth to a drier savanna state and starts belching out carbon rather than storing it? Year on year, we already see this occurring deep in the Amazon. Indeed, some recent research tentatively indicates one-fifth of the Amazon now emits more carbon than it absorbs. Independent of deforestation, the rainforest is storing less and less carbon, as trees die earlier and the carbon they store puffs into the atmosphere as they rot. When you combine this with deforestation, we are coming dangerously close to a tipping point. In fact, Thomas Lovejoy and Carlos Nobre, leading researchers in biodiversity and the Amazon, have suggested that the Amazon could tip if just 20 to 25 per cent of the forest is destroyed.

The Amazon stabilizes Earth's climate. This is why, as 40,000 fires raged through the Amazon in 2019, French president Emmanuel Macron called for the world to protect this priceless global commons.[16] Since 1970, about 17 per cent has gone. We are sitting on a time bomb, but instead of trying to defuse it, we are tapping it with a hammer.

The risks cascade. As one tipping point falls, we lurch towards another, and then another, like falling dominoes. Once we knock the first few dominoes over, it will be too late to keep Earth in a Holocene state. Once we cross a tipping point, cuts to greenhouse gas emissions may make little difference. Earth will release billions of tonnes of gases into the air from permafrost, forests, and under the sea.

The domino effect is scientifically plausible. If it were to begin, then what would be the next stable state of our planet?

In Chapter 1, we introduced the two stable states on Earth: icehouse and hothouse.[17] We have not been in a full hothouse state for at least 5 million years. Back then, carbon dioxide levels were above 350 ppm in the atmosphere, the critical threshold for entering the icehouse. Today, carbon dioxide is already above 415 ppm. Unrestrained fossil fuel emissions may propel Earth back to the

---

16 Global commons, in this context, is a term used to describe the systems that regulate the stability and resilience of Earth.
17 As you will recall, the Holocene, while seemingly stable to us, is not a deeply stable state like the snowball or hothouse Earth or even the long ice ages.

hothouse, crash-landing there in 2200. Luckily, many nations have already committed to deep emissions cuts. If we keep these promises, will the sleeping giants stay asleep? Given all we know about tipping points, we think some concern is warranted. Earth should be safe from an inexorable hothouse Earth fate as long as we manage to keep global warming well below 2°C (3.6°F). Yet, current government commitments take us to 3°C (5.4°F). All bets are off.

For now, our planet remains remarkably resilient. Since we entered the Anthropocene and embarked on this exponential journey, Earth has continued to be our friend. So far, the land and ocean have absorbed half our emissions. However, this cannot go on forever. The capacity of the Amazon rainforest to hold carbon is now decreasing. Similarly, as the ocean warms, it holds less carbon dioxide. Furthermore, the more carbon dioxide the ocean absorbs, the more acidic it becomes.

When we look at the latest science on tipping points and the state of our global commons, it is clear that we have been underestimating the risks facing civilization.

If we succeed in curbing greenhouse gases to attempt to keep Earth's temperature at 2°C (3.6°F) global warming, then Earth's feedbacks may still nudge temperature up to at least 2.5°C (4.5°F). This may be enough to set off the domino effect. The next series of tipping points could push Earth's temperature even higher, and off we go, into an uncontrolled cascade that rewinds the climate clock tens of millions of years in just a couple of centuries.

Sometimes, it takes a generation or more to make incremental progress in science. Earth system science is not like that. When we published the "Hothouse Earth" paper in 2018, we still thought that the world had time to act. It was, after all, a scientific attempt to provide the best possible risk assessment. A year later, our next analysis of tipping points made us hit the panic button.

EMERGENCY ON PLANET EARTH

Our house is on fire.

GRETA THUNBERG
WORLD ECONOMIC FORUM, DAVOS, 2019

It took just three months from the moment the World Health Organization picked up a media statement on cases of "viral pneumonia" in Wuhan, published online by the Wuhan Municipal Health Commission, to the moment nearly 4 billion people – half the population of the planet – were placed under some sort of lockdown. This point was reached on 3 April 2020, according to *The New York Times*.

The COVID-19 pandemic was undoubtedly the worst health crisis in a century. Indeed, it arrived almost precisely 100 years after the Spanish flu pandemic of 1918. The rapid spread of the disease demanded a global emergency response founded on solidarity and cooperation, not an "every man for himself" attitude from nations, states, cities, and companies. The results were mixed.

Countries struggled to contain the virus. Health systems collapsed under the pressure. Forced to act to contain the spread, nations snapped shut large parts of their economies for months on end.

Amid the chaos and confusion, we often glimpsed the best of humanity. Citizens, governments, businesses, schools, and hospitals all showed a remarkable capacity to rise abruptly in the face of an emergency. Many politicians acted responsibly, respectfully, and rapidly. At the height of the crisis, national leaders made strong statements about unity and cooperation and appealed to people's sense of community. And people responded. They made the necessary sacrifices and worked towards the common good.

SLEEPING GIANTS: THE AMAZON

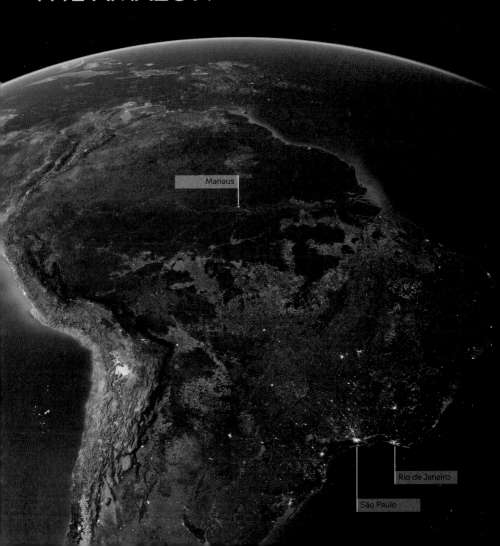

Manaus

Rio de Janeiro

São Paulo

HE GREAT ACCELERATION

SOCIO-ECONOMIC TRENDS

EARTH SYSTEM TRENDS

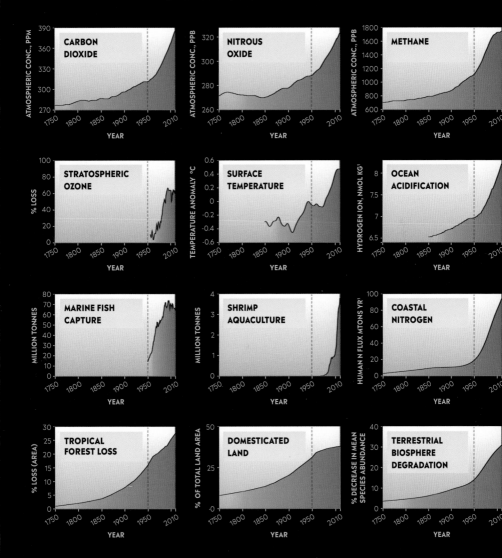

PLANETARY BOUNDARIES

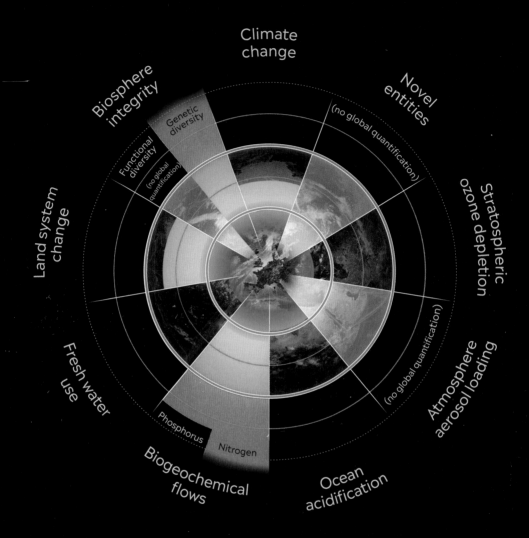

Climate
change

Biosphere
integrity

Genetic
diversity

Functional
diversity
(no global
quantification)

Novel
entities
(no global quantification)

Stratospheric
ozone depletion

Land system
change

Atmosphere
aerosol loading
(no global quantification)

Fresh water
use

Phosphorus   Nitrogen

Biogeochemical
flows

Ocean
acidification

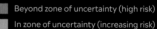

In 2009, researchers identified nine
planetary boundaries. These are the
critical variables for maintaining Earth
in a stable state, similar to the Holocene
– the climate and ecological conditions of
the past 10,000 years. In 2015, updated
research concluded that human impact has
led to the transgression of four planetary
boundaries: climate, biodiversity, land, and
biogeochemical flows (nitrogen and
phosphorus use).

■ Beyond zone of uncertainty (high risk)
■ In zone of uncertainty (increasing risk)
■ Below boundary (safe)
■ Boundary not yet quantified

While many countries acted swiftly and responsibly as scientific knowledge of the catastrophe grew, some national leaders, such as President Donald Trump in the United States and President Jair Bolsonaro in Brazil, dismissed epidemiologists and the advice of other scientists. These leaders did not only question health experts, but also challenged the scientific process and the legitimacy of science to provide guidance. This should give us all great cause for concern as a bigger, more profound crisis engulfs us.

There is a lot to be said, and much will be analysed and debated, about the links between the coronavirus and the planetary scale degradation of our life support systems: the biosphere and the climate system. COVID-19 is a zoonotic disease: one that has leapt from animals to humans. Such interactions have intensified as natural habitats for wildlife disappear and humans live in densely populated cities. There are clear links between biodiversity loss, deforestation, and the emergence of novel viruses. Forest edges are a breeding ground for new strains of human viruses. Research shows that people and their livestock are more likely to come into contact with wildlife when we destroy more than 25 per cent of original forest cover. We can, therefore, make a general conclusion about COVID-19: the pandemic was a result of our connected, crowded, and vulnerable world in the Anthropocene. We cannot hope to prevent future pandemics without also addressing the sustainability of our planet.

In this respect, coronavirus, climate change, and ecological collapse are the same. The key insight is that in the Anthropocene, all of us are not only connected we are also interdependent. In the Anthropocene, what you do affects me, wherever you are or I am on Earth.

The pandemic is a manifestation of our globalized world, characterized by connectivity, scale, surprise, and speed. Now is the time to recognize our failures and vulnerabilities. But we should never let a crisis go to waste. We need to rebuild healthy and sustainable societies that embrace resilience to shocks and human development on a stable planet. To avert our planetary crisis, we need to couple our people's and our planet's health into one transformative agenda for our common future on Earth. Nothing less.

Earth is now unstable. We are losing control. This is an emergency. And scientists are feeling decidedly nervous.

The Intergovernmental Panel on Climate Change (IPCC) is the independent body responsible for marshalling thousands of climate researchers to assess evidence of climate change, risks and impacts of global warming, and pathways to solve the climate crisis. In 2018, based on thousands of scientific papers, it provided strong evidence that approaching 2°C (3.6°F) global warming would lead to unacceptable suffering and economic hardship for hundreds of millions of people across the world. As a result, 1.5°C (2.7°F) global warming is now scientifically established as the climate planetary boundary. It is not a safe level; let us make that very clear. There are still risks. But it is *significantly* (that is a big word in science) safer than 2°C (3.6°F). Beyond 1.5°C (2.7°F), economic and social costs rise fast and in ways that we do not yet fully understand.

At global warming of 1.5°C (2.7°F), maintaining crop yields, managing disease spread, and dealing with heatwaves, droughts, and other extreme weather will not be easy. But it will be far easier than at 2°C (3.6°F). This is becoming increasingly obvious. Nudging the mercury up a further 0.5°C (0.9°F) will be an unpleasant, terrifying journey into the unknown. The United Nation's Paris Agreement of 2015 commits countries to act to keep temperature well below 2°C (3.6°F) and to aim for 1.5°C (2.7°F). The science confirms that this is a sound decision for our future on Earth. However, the actions that countries have promised to take are only enough to hold temperature to 3°C (5.4°F). And, as we explained in Chapter 7, we cannot rule out that going beyond 2°C (3.6°F) may set off a cascade of tipping points, sending us towards a hothouse state.

So, in summary, the key insights we truly need to understand are the following. First, overwhelming science now clearly shows that holding the 1.5°C (2.7°F) planetary boundary line will give us humans a manageable future. Go to 2°C (3.6°F) and impacts will be more ferocious. Second, this future only considers immediate impacts. We also have the risk of crossing tipping points, triggering self-reinforcing warming by Earth itself. Pass 2°C (3.6°F) and we cannot exclude that we may be pushing the "on" button, raising global temperatures another 0.5°C (0.9°F). This then risks an unstoppable cascade. The journey towards a hothouse would start.

In addition to this climate-related analysis, in 2019 the most important scientific report on biodiversity was published. The Intergovernmental Science-Policy Platform on Biodiversity and Ecosystem Services global assessment concluded that 1 million species risk extinction.

Is all of this enough to declare a planetary emergency?

Declaring a planetary emergency

Deciding to declare a planetary emergency cannot be taken lightly. But who can declare such an emergency? Who has the authority? The responsibility falls to political leaders. The role of scientists is to help determine whether there is enough evidence to do so. For example, during the Chernobyl (1986) and Fukushima (2011) nuclear disasters, politicians and engineers needed to understand the immediate and longer-term risks so that they could make the right decisions. With expert advice from scientists, they were able to act quickly to contain the catastrophes in what were clearly emergency situations.

In 2019, we explored with scientific colleagues whether the case for a planetary emergency was scientifically justified. Spoiler alert: it most definitely is.

We came to this conclusion based on two lines of reasoning. First, the world needs to stay as far below 2°C (3.6°F) global warming as possible. The best way to do this is to exit fossil fuels immediately, stop deforestation, and rebuild the resilience of forests, oceans, and wetlands. Realistically, we probably will not be able to keep temperature below 1.5°C (2.7°F). We can only emit a further 320 billion tonnes (350 billion tons) of carbon dioxide if we want a two-in-three chance of reaching this target. Currently, we emit about 40 billion tonnes (44 billion tons) a year. So, each year, we rip through more than 10 per cent of our remaining carbon dioxide budget.

The second line of reasoning relates to tipping points. When the first major research on tipping points was published in 2008, there were no signs that we might be getting close to waking the sleeping giants. Back then, we thought that there would only be a high risk of this happening at 3 or 4°C (5.4 or 7.2°F) global warming.

In 2019, we revisited this research. What we discovered was the biggest shock of our careers. As we pored over the data, we found

Active tipping points

Until very recently, scientists thought the dangers of crossing tipping points would arrive much later this century. Now, we can measure large-scale changes already under way in many important places.

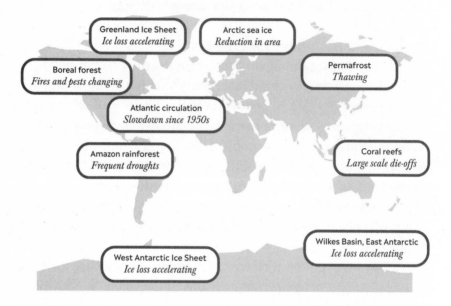

Greenland Ice Sheet
Ice loss accelerating

Arctic sea ice
Reduction in area

Boreal forest
Fires and pests changing

Permafrost
Thawing

Atlantic circulation
Slowdown since 1950s

Amazon rainforest
Frequent droughts

Coral reefs
Large scale die-offs

Wilkes Basin, East Antarctic
Ice loss accelerating

West Antarctic Ice Sheet
Ice loss accelerating

that many of the cascading tipping points are already at risk and are beginning to change in profound and, frankly, quite disturbing ways.

The permafrost in Siberia is now thawing. Russians have built roads, houses, factories, and pipelines on permafrost, believing it to be, well, permanent. Roads and pipelines are buckling. Permafrost is not meant to behave in this way. But the bigger risk, as always, is the greenhouse gases trapped in the ice, sometimes known as a carbon bomb. Permafrost in the Arctic region holds 1.7 trillion tonnes (1.87 trillion tons) of carbon. This is more than double the carbon we have emitted into the atmosphere from fossil fuel burning since the start of the industrial revolution. This is stable as long as the permafrost remains frozen. But the Arctic is the fastest-warming region on Earth, and it is destabilizing. What happens if it starts to accelerate? Unfortunately, this is not the only risk in the region. The forest and peatlands of the far north are drying out, too, creating bigger, more fierce fires in the Arctic. In 2019, the smoke cloud billowing from Siberian forest fires was larger than Europe.

Greenland has lost almost 4 trillion tonnes (4.4 trillion tons) of ice since 1992. Ice is slipping into the sea at an accelerating pace. During 2019's heatwave, 600 billion tonnes (660 billion tons) of ice melted in two months, raising sea levels 1 millimetre (0.04 inches) per month. Greenland's tipping point was once estimated at above 2°C (3.6°F) global warming, but now we have to seriously ask, could it reach a tipping point below 2°C (3.6°F)?

And in 2020, researchers measuring how the Amazon rainforest is changing published a devastating report showing that the world's most important tropical forest is on its knees. Year on year, it is storing less and less carbon. It can no longer cope with our attacks. By 2035, and perhaps as early as 2030 – in just a decade – the rainforest may switch from being a store of carbon to a major emitter. We are in deep trouble.

But perhaps most shocking are the changes to Antarctica.

We used to think that Antarctica was relatively stable, certainly more stable than Greenland, and that we would only hit the high-risk zone if global temperatures rose well beyond 2°C (3.6°F). We have known for some time that the West Antarctic Ice Sheet – the spiky tail of the continent towards South America – was potentially unstable, as global temperatures rise. The East Antarctic Ice Sheet, on the other hand, was thought to be very resilient to warming: that is, Earth would need to reach a scorching 5°C (9°F) warmer than today to budge it. All that has now changed. In fact, nine of the 15 known major tipping elements of the Earth system are on the move, showing signs of weakening.

In 2019, researchers found a cavern two-thirds the size of Manhattan beneath Thwaites Glacier in West Antarctica. Melting seems to have hollowed out this vast cavity over the past three years. A recent expedition found that the water temperature at the grounding line – the very point a glacier loses contact with the rock beneath and becomes a floating ice shelf – is at 2°C (3.6°F) above freezing. This is bad news. There is nothing stopping the retreat of the grounding line ... and the glacier.

How bad is this news? The Thwaites and Pine Island Glaciers protect the whole West Antarctic Ice Sheet from collapse. They function as plugs holding the upstream glaciers from sliding into the ocean. If the whole ice sheet disintegrates, the water it releases will be enough to raise sea levels by 3 metres (10 feet). This could take a few centuries according to some models, or it could be much faster. We cannot rule out the collapse of Thwaites Glacier within decades. The reason the ice sheet is likely to go is because it rests in a dip below sea level. The grounding line where ice meets water meets rock has now crept over the rim of the dip, and the water can flood unimpeded under the glacier, thereby accelerating collapse. Once you see the inevitability of the physics, you start to get extremely nervous indeed.

In 2019, the authors of the IPCC blockbuster report on the ocean and ice sheets concluded that parts of East Antarctica now look to be destabilizing, too. When scientists detected the hole in the ozone back in the 1980s, we were shocked. We saw the risks. And we acted. Now, we have woken a second sleeping giant at the South Pole. We are shocked. We see the risks. And we must act. Even if we have crossed the point of no return, there is still a chance we can control the speed of collapse and probably prevent the entire collapse of Antarctica (which holds enough water to raise sea levels more than 60 metres [197 feet]).

Here's what we know. Significant parts of Antarctica and Greenland are destabilizing. Computer models say this is possible at current temperatures, the data show they most definitely are, and geological data from ice cores and rocks indicate that in the past they have destabilized at the same temperature. That is three separate lines of inquiry. As Douglas Adams said, "If it looks like a duck, and quacks like a duck, we have at least to consider the possibility that we have a small aquatic bird of the family Anatidae on our hands."

The paleo data are interesting. We can look at past interglacials during the ice age cycles of the past 2.7 million years. When temperature rise was a little lower than today (1°C [1.8°F] global warming), sea levels stabilized at 6 to 9 metres (20 to 30 feet) higher than today, as a result of a destabilized Greenland and Antarctica. When temperature was 1 to 2°C (1.8 to 3.6°F) global warming, sea levels rose up to 13 metres (43 feet) above today's levels. In this scenario, most coastal cities will have to adapt or die. The costs will be colossal. Our actions now determine the future of these cities, and of the world's small islands and low-lying coastal areas.

So, in summary, there are two types of tipping point risk, and currently we appear to be embracing both enthusiastically. First, ice sheet collapse. Second, carbon bombs: turning stable carbon storage systems – the rainforests, boreal forests, and permafrost – into unstable carbon sources. We are juggling fireballs while standing unnervingly close to a gas station.

Risk and urgency

The transition from stable planet to unstable planet is happening on our watch. We have little time to act, and we risk irreversible catastrophe, affecting all future generations, if we do not respond swiftly. The two-pronged rationale for issuing a planetary emergency is clear. An emergency is related to risk and urgency. We have already established that the risks are now sky high. When it comes to urgency, we have 10 years to cut emissions in half, at the very least, and 30 years to become carbon neutral. It will take a minimum of 30 years to achieve this. If a patient suffers a heart attack and needs to be treated within five minutes, the situation is controlled if paramedics can get there within that time. It is out of control if they cannot. We are now at risk of spinning out of control.

Declaring an emergency and ramping up action right now also make sense for the economy. When COVID-19 took a hold on our planet, economies went into shock. This made managing the crisis doubly difficult. An economy in free fall is hardly a good foundation for transformative action. Acting on the planetary emergency now and focusing on using the economy to drive societal transformation will combine growth in prosperity with resilience.

At the time of writing, more than 1,700 towns, cities, councils, and regions have declared climate emergencies in 26 countries, covering almost 1 billion people. Indeed, the movement to declare emergencies has been rising exponentially. On 1 May 2019, the UK parliament declared a climate emergency. One month later, during a chaotic period in UK politics, as Brexit dominated headlines and political parties could barely agree on anything, politicians did manage to find some common ground: parliament agreed to enshrine a hugely ambitious climate target of reaching net zero by

2050. The United Kingdom is the first major economy to do so. Personally, we were stunned by the speed and conviction shown. This was utterly unthinkable just 12 months earlier.

Humanity has faced myriad disasters before. In fact, a remarkable trait of human resilience and ingenuity is our extraordinary stamina and capacity to rise after disasters. One of the first recorded natural catastrophes is a colossal flood that appears in several mythologies and religions, most notably in the biblical story of Noah's ark. Given this catastrophic flood features in the writings of several civilizations, could it be based on a real event? Some scholars think it may be a direct reference to the major global flooding that must have occurred at the end of the last ice age, when some ice sheets abruptly collapsed and the sea level rose many metres within a short period of time. Word of mouth passing from grandparents to grandchildren could plausibly have kept the tale alive for 200 generations, until the dawn of cuneiform and stone tablets.

We have not had to contend with epic floods for 10,000 years, as we have been blessed by the relative stability of the Holocene. As we step deeper into the Anthropocene, though, such apocalyptic flooding becomes increasingly plausible, if not inevitable. So, too, does the spread of more disease, affecting humans, animals, and plants. Without action, we risk the collapse of grain-producing regions through extended droughts. And a spike in heat-related deaths. And growing carbon bubbles and market volatility. We are at the point of no return. This is not a drill.

Consequently, we have, for the first (and probably only) time in modern history, enough evidence to declare a planetary emergency.

The Earthshot

In 2019, we collaborated with Sandrine Dixson-Declève, the co-president of the Club of Rome, Bernadette Fischler, head of advocacy at WWF-UK, and other colleagues to help draft a planetary emergency declaration. This was launched at a United Nations event in New York, linked to the United Nations Climate Action Summit. More than 10 heads of state supported the call for the declaration, including UK Prime Minister Boris Johnson.

In it, we looked to leaders like Johnson to acknowledge humanity's greatest existential threat. In his own inimitable style, Johnson opened the event saying, "There are more UN bureaucrats in the world than there are Indian tigers on Earth. There are more heads of state than there are humpback whales." Although they were once heading to extinction, humpback whales now number more than 100,000 – a success story of government intervention. But we know what he was trying to say.

The planetary emergency declaration identified the leverage points for systems change within the global economy:

- Issue global moratoriums on deforestation and further fossil fuel exploration.
- Stop further expansion of agricultural land, and build up the resilience of our most vital ecosystems.
- End the crazy subsidies for fossil fuels, totalling about USD 500 billion every year, and more like USD 5 trillion when you count health impacts and pollution. In fact, if fossil fuels were priced correctly, greenhouse gases would be reduced by more than 28 per cent.
- And put a price on carbon starting at a minimum of USD 30 per tonne to shift the markets.

These actions are needed immediately. In total we identified 10 actions to protect the global commons and 10 actions to drive economic and social transformation. We updated the text in 2020, in light of the COVID-19 pandemic, to emphasize actions that reduce the risk of further global health crises and that align the economic recovery from the pandemic with the goal of restabilizing Earth. This is the bare minimum needed to send strong signals to markets to ensure they remain stable during the transformation. In reality, this is just the beginning.

As we come to a close on Act II, the stakes could not be higher. But we have the beginnings of a plan. In Act III, we will further develop the Earthshot.

The starting point is a new world view: true planetary stewardship – operating the global economy within planetary boundaries.

ACT
III

PLANETARY STEWARDSHIP

Nature is not a sector of our society, it's a prerequisite for it. We humans do not exist outside the biosphere, we're part of it. We are dependent on the biosphere and our actions – with their outsized scope and speed – have an impact on the planet's capacity to sustain us. We have to reconnect with our planet.

CARL FOLKE
STOCKHOLM RESILIENCE CENTRE

The world is changing. Take Adam Arnesson, a farmer in central Sweden. As a farmer, he produces milk. But he has given up his dairy herd and now grows oats instead. He produces more food than ever before, and his farm is more profitable. Researchers have found that swapping cows for oats can reduce greenhouse gas emissions by up to 41 per cent. By the way, Adam no longer calls himself a farmer; he says he is now a biosphere steward.

Demand for plant-based milk is insatiable. Since 2010, the market for milk in the United States declined, but oat milk sales grew 686 per cent in 2019. The popularity of plant-based milks is pushing companies across another tipping point: bankruptcies in the dairy industry. We may be seeing a new equilibrium emerging, with sustainability and health guiding both cow-based and planet-based dairy production.

Indeed, more people like Adam are adopting new ways of thinking about our world. Farmers are noticing how the buzzing bees and chirping birds that once accompanied their work have now gone quiet, and that the streams and rivers around their lands have become dead zones. Many are thinking beyond their own horizons, deepening their knowledge of the biosphere. They are discovering how each patch of land is woven into an ecological tapestry. If they pull the threads too hard in one place, the whole tapestry unravels.

This interconnectedness is not confined to farming. We have seen how a virus spillover, probably originating in Wuhan, China, can create havoc across Earth. Losing a rainforest can change weather patterns, leading to more severe droughts in other countries and making it harder to control greenhouse gas emissions. Protection from disease is a precious global commons, as is protecting the resilience of our large ecosystems on land and in the ocean. Collective action can stop an infection reaching millions of people, as seen to powerful effect in places like New Zealand, Greece, and South Korea during the pandemic. Collective action to live within planetary boundaries can halt risks of environmental collapse. Together, such action will provide us with pathways to better, more attractive, secure, fair, and modern lives. For everyone.

We call this planetary stewardship.

The Anthropocene forces us to reassess our relationship with Earth. We need to change our world view and we need a plan. If we do not alter how we think and act, we are on course for "the uninhabitable Earth" predicted by the US writer David Wallace-Wells in his book of the same name. If planetary stewardship is our new guiding philosophy, our North Star, then our plan and our mission this decade, and indeed for several decades to come, is the Earthshot, which will direct the world's economic engine in order to restabilize our life support system. The goal is nothing less than the long-term habitability of our planet. In this chapter, we will explore how planetary stewardship is emerging all around us; we will redefine the global commons in line with our latest understanding of the Anthropocene and discuss a new plan, the Earthshot.

Planetary stewardship is a paradigm shift in how to think about the world. In 2000, when Paul Crutzen declared that we were no longer in the Holocene, but rather in the Anthropocene, this marked a scientific paradigm shift. Until that moment, scientists were locked in the old way of thinking about the planet – a Holocene world view. But paradigm shifts are not the exclusive preserve of science. They also matter in the real world. We live on a finite planet, but we are acting as if the planet is infinite. Our material use is rising exponentially; so, too, is our waste. The lily pond may have seemed infinite just one or two decades ago, but now we are rubbing up against the boundaries. If we are to survive this century, we need to update our world view.

Back in 1972, a group of complex systems researchers published a landmark analysis titled "Limits to Growth", commissioned by the influential Club of Rome. The analysis caused a stir, and it still does to this day. It concluded, "If the present growth trends in world population, industrialization, pollution, food production, and resource depletion continue unchanged, the limits to growth on this planet will be reached sometime within the next one hundred years. The most probable result will be a rather sudden and uncontrollable decline in both population and industrial capacity." Since 1972, the world has largely continued on the same course of exponential resource extraction and pollution. The "Limits to Growth" analysis indicates that we should be seeing some signs of major ecological, economic, and social problems between 2015 and 2030. Few would doubt that this is precisely what we are witnessing right now. In fact, updates of the original study in recent years show that the world is largely following the risky paths predicted by the original analysis.

One of the original researchers on the "Limits to Growth" report, the late systems scientist Donella Meadows, became fixated with how complex systems like societies change course. At one particular meeting on global trade, participants were discussing how to connect two institutions: the World Trade Organization and the North American Free Trade Agreement. While listening to the discussion, it occurred to Meadows that the participants were not thinking in the right way about changing systems. She recalled, "Without quite knowing what was happening, I got up, marched to the flip chart, tossed over a clean page, and wrote."

She scribbled down nine places in which to intervene in a complex system, in increasing order of effectiveness (see overleaf). The last,

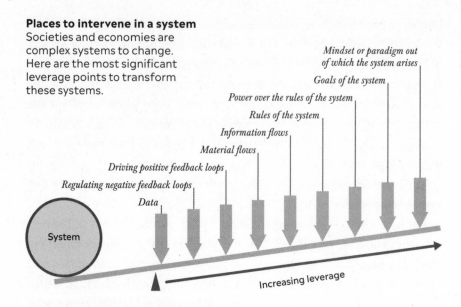

Places to intervene in a system
Societies and economies are complex systems to change. Here are the most significant leverage points to transform these systems.

Mindset or paradigm out of which the system arises
Goals of the system
Power over the rules of the system
Rules of the system
Information flows
Material flows
Driving positive feedback loops
Regulating negative feedback loops
Data

System

Increasing leverage

and most effective, place was "The mindset or paradigm out of which the system – its goals, power structure, rules, its culture – arises." "Everyone in the meeting blinked in surprise, including me," she later said.

In our global intertwined system of 7.8 billion inhabitants living within a complex biosphere, the best way to change course is to alter the lens through which those people in the system view the world. Our friend and colleague Carl Folke is the world's foremost thinker on sustainability and resilience. He starts almost every single one of his talks with the most profound yet basic observation, "We must reconnect with the biosphere: the living part of the planet." This is the essence of planetary stewardship.

This may seem obvious. It is like saying, "Hey guys, remember, we live on a planet and we depend on it being stable." As if we had forgotten. But when you step back a bit – while driving down asphalt roads surrounded by concrete, steel, and glass, on your way to the shopping mall, to fill up on basic goods like food, and materials for shelter, safety, and comfort – you must admit that, yes, most of us have disconnected from the planet. A slow, silent, but all-encompassing disconnect. We take our planet for granted, at least its stability. Perhaps this is understandable: we have, after all, always lived on a stable planet, and a seemingly infinite one at that.

But climate extremes are beginning to shake us from our slumber. According to the United Nations Environment Programme, by the close of 2019, Brazil, the Democratic Republic of the Congo, Russia, and the United States had all experienced megafires (fires greater than 400 square kilometres or 100,000 acres) on what many have called "unprecedented scales". And then the infernos hit Australia in late 2019 and early 2020, edging into the suburbs of major cities. Economic losses from weather and climate-related natural disasters are increasing worldwide. And it is always the poorest who are hit hardest, even in the wealthiest countries. A paradigm shift towards planetary stewardship, as an integral part of human development, is today a very necessary next evolutionary step in our world.

Planetary stewardship is hardly a new idea

Let's stop and take a deep breath. Native American and indigenous peoples have always talked about environmental stewardship and were practising it long before the cowboys, trappers, and homesteaders appeared on the horizon; it is not a new concept.

We have been privileged to be part of the Edmund Hillary Fellowship and Institute in New Zealand, which aim to connect sustainability researchers with innovators and Māori culture, and more broadly New Zealand/Aotearoa culture. It has been humbling to see how our work and the way of thinking that we are trying to promote are actually not so novel. Stewardship is embedded in Māori culture, from the very first greeting, which goes something like "I am not here representing myself; I am representing all my ancestors", to words such as "kaitiakitanga", meaning conservation, replenishment, or sustainability of the environment.

When notions of environmental stewardship clash with the modern world, interesting solutions emerge. In 2017, the New Zealand/Aotearoa government passed a new law granting the Whanganui river the same rights, powers, and duties as a legal person. The practical upshot of this is that the river can sue you if you mess with it. This sounds extraordinary, until you think that governments have been granting companies and corporations legal personhood for centuries. (It is quite interesting that newspapers refer to corporations as living organisms: "Microsoft announced today it would open the doors to its storehouse of patents ... .") The river is not alone. A former national park, Te Urewera, and a mountain, Mount Taranaki, have also now gained legal status.

The Māori world view, or *te ao Māori*, is deeply connected to environmental stewardship. This world view recognizes long-time horizons: a responsibility to distant ancestors and future generations. Te ao Māori articulates a profound knowledge of the interconnections between ecosystems and societies. The Ngāti Wai and Ngāti Whatua communities, for example, have a saying, "Ko ahau te taiao, ko te taiao, ko ahau" – meaning "The ecosystem defines my quality of life." Concepts related to deep time and resilience are captured through ideas such as "Mō tātou, ā, mō kā uri ā muri ake nei" – "For us and our children after us." The Māori have a knowledge system that they call Mātauranga Māori. It describes their approach to environmental management, policy development, and implementation. Sometimes Mātauranga Māori is referred to as "ancestral knowledge". But this implies some sort of dogmatic approach to knowledge and fails to capture how this knowledge evolves over time. Although there is no direct translation between Mātauranga Māori and scientifically defined concepts of stewardship, there are striking similarities between the underlying philosophies of Mātauranga Māori and planetary stewardship.

The Maori are not alone. Similar concepts are embraced by many other indigenous cultures. In 2012, the Bolivian government passed the Law of the Rights of Mother Earth, granting nature the same rights as humans. Ecuador has introduced a similar law. And here in Sweden, we have the word "lagom", which means "not too much, not too little". This short word is so deeply embedded in our culture that it helps define the Swedish world view.

These ideas challenge the dominant world view of industrialized societies that emerged over the past two centuries based on the nation state narrative: "Your country needs you to consume more." Waste and environmental pollution are viewed as unavoidable yet manageable consequences of production and consumption.

Redefining the global commons
The US political scientist and Nobel Laureate Elinor Ostrom spent her career looking at how communities manage common resources, such as local forests, pastures, and fishing grounds. She

showed that communities can manage resources effectively for many generations, often with little or weak government regulation, and identified eight design principles that allow this.

Towards the end of her career, she started looking at the global commons. In 1999, Ostrom and her colleagues proposed how to adapt the eight principles for planetary stewardship. They recognized that for the stewardship of an ecosystem, that ecosystem must continue to be useful. Exploitation cannot be so complete as to have drained all the resources. A fishing ground with no fish has no value. On the other hand, the resource cannot be so unscathed that there are no obvious benefits to managing it.

Ostrom's first principle for stewardship is to define clear boundaries for resource use. For the first time, the planetary boundaries described in Chapter 6 fill this criterion at the global scale. Other principles state that all users must have access to accurate knowledge of the resource conditions, and people who use a resource, like the ocean or forest, must see how the benefits of protecting that resource outweigh the cost of doing so. This means that the group benefits and that the costs are not just personal.

We have now reached a turning point for planetary stewardship. The conditions Ostrom outlines are precisely where we are right now. The biosphere has not yet been drained so completely that it has passed a point of no return, although its resilience is weakening. With the planetary boundaries framework, we now have a first, rough estimate of the nine boundaries that ensure sustained resilience into the future. And increasingly, through a planetary network of satellites and other monitoring systems, we have accurate information on who is using the resources and how much they are taking. A system for planetary stewardship is most definitely emerging.

For the past century, the global commons have been recognized in a narrow legal sense to manage four zones beyond national jurisdiction: Antarctica, the High Seas, outer space, and the atmosphere. In 2015, we were invited to write a report about what the global commons means in the Anthropocene. Influenced by Ostrom's work, we published a report with two colleagues, Nebojsa Nakicenovic and Caroline Zimm, that proposed a new definition of the global commons. We argued that all parts of the Earth system that protect Earth's ability to remain in a Holocene-like stability –

the planetary boundaries – must now be viewed as global commons. Every child's birthright is, after all, a stable and resilient planet. When we see children walking out of their lessons to join the school strikes, we know precisely what they are fighting for: to restabilize Earth.

This is a challenging idea for a planet organized by nation states. But our analysis only accounts for the physical and biological reality of the planet, not the geopolitical reality of nation states. When French President Emmanuel Macron offered international help to control Brazil's 40,000 forest fires in 2019, Brazilian President Jair Bolsonaro rejected the offer. Why should anyone dictate to Brazil what it can do with its rainforest? But what would happen if a nation attempted to blow up the Moon? There would be outrage. The Moon influences the tides on Earth – a global common – and light at night, the length of the day, and a whole host of other things. The same applies to the Amazon: killing the rainforest will affect everyone. We need to find new ways of governing in order to stabilize Earth because business as usual will drive us off the cliff. In short, we can put all this together in a simple equation:

Planetary boundaries + Global commons = Planetary stewardship

Has planetary stewardship become a thing?

As we have seen, stewardship as a world view has a long history. But is planetary stewardship now catching on? Books such as Edward O. Wilson's *Half-Earth: Our Planet's Fight for Life* (2016) encourage a different relationship with nature, where ecosystems are given space to thrive and connect. David Attenborough's Netflix series *Our Planet* (2019) connects with profound ideas around stewardship at an Earthly scale. Perhaps a tangible shift is now under way in unexpected places: planetary stewardship is no longer an occasional lone voice, it is a global movement across economic sectors. Our children are reading scientific papers and walking out of schools. The chairman of the Financial Stability Board set up after the financial crisis of 2008, Mark Carney, declared climate to be a systemic risk to the global financial system. In 2020, Larry Fink, CEO of BlackRock, the world's largest investment firm, warned companies, "We will be increasingly disposed to vote against management and board directors when companies are not making sufficient progress on sustainability related disclosures and the business practices and plans underlying them." And we see the leading seafood companies meeting scientists to make plans to become sustainable ocean stewards (Seafood Business for Ocean Stewardship, SeaBOS).

Perhaps the most profound shift, though, is the commitment from all nations to the 17 Sustainable Development Goals (see plate D1) set by the United Nations to be achieved by 2030. These cover everything from ending poverty and hunger on Earth to protecting the biosphere. It is not only countries that have committed to them. Of 730 multinational companies, more than 70 per cent mention these goals in corporate reporting, and almost 30 per cent include them in business strategy.

Science fiction author William Gibson once told *The Economist* newspaper, "The future is already here – it's just not evenly distributed." We would update this slightly to, "Planetary stewardship is already here, it's just not evenly distributed."

We still have a mountain to climb for planetary stewardship to go mainstream, because the scale of the planet brings unique challenges. There is often a huge disconnect between our individual action and our impact. Our information flows are faulty. As our colleagues at the Stockholm Resilience Centre, Beatrice Crona and Henrik Österblom, point out, the economic system relies on price to transmit information. If a fishing ground is depleted, prices should rise as the resource becomes more scarce. But this does not happen. When we buy fish in the supermarket, we largely see stable prices, even though much of the world's fish stocks are at their limit or close to it. Why is this? As one fishing area is depleted, boats steam to the next zone – using cheap, generally subsidized, fuel – and fish there, using cheap labour. We keep depleting one area and then moving farther afield, until fish once caught off shore are frozen and shipped from the poles. No strong price signal will appear until all fishing zones are exhausted. The markets do not account for the Anthropocene.

The economic system is not our only source of information. Even with the power of the Internet and instant access to much of the world's information, the signal to noise ratio is atrocious. Reliable information is not mainlined into our brains; it joins an information deluge. The technology companies that have built their reputations on organizing the world's information (Hello Google) or building a global community (Hi Facebook) have created vast toxic clouds of confusion. Some tweaks to some algorithms need to be made here before we can say planetary stewardship has arrived.

The Earthshot mission

This chapter has been about resetting and updating world views. Resetting world views is not easy, though. We are tribal societies. We adopt information that supports our tribal world view, and we reject information that challenges it, even if that means ignoring the best available science. We can see this behaviour in the information wars around vaccinations, nuclear power, genetically modified food, and even evolution. What the scientific evidence is demanding now is profound. The scale of the mission is far greater than a Moon landing or banning ozone-destroying chemicals. Science is indicating that the economic foundation of the modern world needs to change utterly within a decade or two. It is no surprise this message is met with hostility and anger.

So, how do we act on the latest science? Very practically, what do we need to do to stabilize Earth? And how do we do it fast enough and at large enough scale, despite all the social and environmental complexities? There are some obvious places to start: controlling greenhouse gases, changing how we produce food, stabilizing population growth. Together with our colleagues working on The World in 2050 project, we have identified six system transformations that need to happen in the next decade to slow the rate of change of Earth's life support system and to allow all a chance of a good life on a stable and resilient planet. In operational terms, this means meeting the Sustainable Development Goals within planetary boundaries, not only by 2030, but also maintaining the good work to 2050 and beyond. This is the Earthshot mission, which we will explain further in Chapters 10 to 15. We will also discuss, in Chapter 18, the four tipping points – social, political, economic, and technological – that will drive these transformations. We understand that such transformations can be overwhelming; they are for us, too. We stay optimistic, though, because these transformations are not starting from scratch; they have been building for 50 years.

Progress has been so slow that it looks incremental. This is frustrating and makes many think we have made no progress at all. In fact, we are at the knee of the exponential curve. Things are about to take off in the 2020s.

How might this happen? We are not claiming that everything we describe will fall magically into place exactly as we suggest. Undoubtedly, it will be messy. But Bill Gates once said that we overestimate what can be achieved in one year, but underestimate what is possible in 10 years. This gives us hope. To illustrate this point, here are some examples.

In 1961, US President John F. Kennedy announced the goal of landing a man on the Moon by the end of the decade and channelled 2.5 per cent of the country's gross domestic product (GDP) to make it happen. In 1969, this goal was achieved.

After the discovery of the hole in the ozone layer, countries had to act fast in order to avoid catastrophe. Between 1988 and 1998, emissions of ozone-depleting chemicals fell by 57 per cent worldwide, and they have declined a staggering 98 per cent since 1986.

And between 2007 and 2017, deaths from HIV/AIDS fell by about 50 per cent.

The world harnessed economic power to achieve amazing things. But where will we be 10 years from now? First, we need to put the foundations of the Earthshot mission in place – the six system transformations: energy, food, inequality, cities, population and health, and technology. From the evidence at hand, if we are successful, we have a good chance of winning the grand prize: meeting the Sustainable Development Goals within planetary boundaries.

THE ENERGY TRANSITION

Everybody asks me, what is the biggest threat to climate change? Short termism. That's the biggest threat.

CHRISTIANA FIGUERES
FORMER EXECUTIVE SECRETARY OF THE UNITED NATIONS
FRAMEWORK CONVENTION ON CLIMATE CHANGE AND
FOUNDING PARTNER OF GLOBAL OPTIMISM

Climate change is not something we will ever really solve. We will be managing the stocks and flows of carbon in the atmosphere, ocean, land, and life for the rest of humanity's time on Earth, at least as an industrialized civilization. This is a new responsibility for our species in the Anthropocene.

Ultimately, though, we need to exit the fossil fuel economy. We must slash emissions 50 per cent by 2030, then slash them by a further 50 per cent by 2040, and so on. At the very least. Globally. We have called this pathway the Carbon Law, and it translates to a fall in emissions of about 7.5 per cent every year. While this may be technically possible, it will not be easy. As the world witnessed during the COVID-19 pandemic, greenhouse gas emissions fall dramatically when economies are deeply destabilized. But can emissions fall even further, while retaining economic stability, driving employment, and delivering prosperity? After the Paris Agreement in 2015, we started thinking about how to frame this mammoth climate challenge in a way that communicates the scale, urgency, and opportunity.

The Earthshot's first system transformation is energy. People wonder why so little has been done for so long. We have to acknowledge that the reason for the shockingly slow progress is the nature of what science is demanding: a complete reconfiguration of the foundation of the global economy – our energy system. This should not be undertaken lightly. We should not be surprised by fierce resistance. But we have reached the end game for fossil fuels. This is one of the two system changes that are now locked in: the world is leaving the fossil fuel era. (The other unstoppable juggernaut is the technological revolution.) The big question is not whether we will phase out fossil fuels; the issue is whether the energy transition will happen fast enough.

Our story of the energy transition begins in a Paris suburb in December 2015. After two exhausting weeks – and two decades of failure – the final negotiations of what has become known as the United Nations Paris Agreement on Climate began at 5pm on 12 December, a day later than planned. In the cavernous chambers, expectations were muted. No one seemed to know how the evening would unfold. Based on previous experience, we predicted a confusing, drawn-out night, during which weary negotiators would crawl towards a resolution, fighting tooth and nail over every asterisk, word, and clause.

But it was not to be. At 7:16pm precisely, French foreign minister Laurent Fabius strode onto the stage and announced all issues were now resolved. Before anyone could really process what he had said, he brought down the gavel on the table with a bang. The 21st session of the Conference of the Parties (COP 21) ended abruptly and spectacularly. After a beat, the assembled throng erupted in cheers and applause. We had made it. We had crossed the line. After decades of soul-destroying failure, we now had a global deal on climate. It seemed too good to be true.

Christiana Figueres, who masterminded the conference as head of the United Nations Framework Convention on Climate Change, proclaimed, "We have made history together. Successive generations will, I am sure, mark 12 December 2015 as a date when cooperation, vision, responsibility, a shared humanity, and a care for our world took centre stage." While far from perfect, the deal was much better than we had anticipated before going into the negotiations.

For more detail, let's rewind to the day before. It is midday on 11 December. With colleagues, we hastily convened a press conference for the scientific community to voice serious concerns about the deal on the table. Throughout the previous 14 days, negotiations had gone better than expected. On 9 December, the latest text had landed. Incredibly, it included an aim to reduce fossil fuel emissions by up to 95 per cent by 2050. The draft even mentioned aviation and shipping emissions, which are normally ignored because they are too controversial (a massive failing in previous negotiations). The text committed to follow the best available science. At this point, we were cautiously optimistic that the Paris Agreement might actually work.

Then the penultimate text arrived the following day. The negotiators had ripped up critical parts of the deal. Gone were the clauses axing aviation and shipping emissions. Even the commitment to follow the best science had been removed. Kevin Anderson, a climate scientist and firebrand from the Tyndall Centre for Climate Change Research, recalls, "There was a real sense of unease among many scientists present." What we read did not match the boundless optimism in the corridors, halls, and news reports.

We felt the scientific community simply had to make a stand to say clearly and unequivocally that the text as it stood risked catastrophe. Owen and Denise Young (another firebrand) announced a press conference for the following day. We immediately came under fire from all sides. Lobby groups and even other academics criticized us and leaned on us to cancel the press conference. As Anderson puts it, "Desperate to maintain order, the rottweilers and even their influential handlers threatened ... those daring to make informed comment." Right up to the moment that the microphones went live, our phones buzzed with dire warnings that we were naively bulldozing through the most complex, delicate international treaty ever negotiated. We were blatantly asked, did we really want to be responsible for destroying the Paris negotiations? We stuck to our guns, though. Not least because if the scientific community really held so much sway, the outcome would be based on the best evidence: logically, the agreement should be better, not worse.

Under the banners of the International Council for Science, Future Earth, and The Earth League, we assembled in a room normally allocated for side events. (The United Nations could not find us an official press conference room.) Johan was joined on stage by Hans Joachim Schellnhuber, Joeri Rogelj, Kevin Anderson, and Steffen Kallbekken, with Denise Young chairing. Journalists started streaming in. Soon, all seats were taken, and still they kept coming. People were sitting cross-legged on the floor, leaning along the walls, and, eventually, perching on the edge of the stage. A security guard warned that we were breaking every regulation and demanded we stop. We politely negotiated an amnesty.

The atmosphere was electric. Anderson declared that the deal now on the table was far worse than the woeful deal scrambled together at the ill-fated Copenhagen summit of 2009. The room of hardened, cynical journalists gasped in astonishment. For the first time, people were publicly challenging the carefully crafted media and political narrative of unquestioned optimism that had taken hold over the previous weeks. We argued that while the text maintained an aspirational 1.5°C (2.7°F) global warming target, the agreement lacked teeth. Ships and planes had been shunted under the carpet and ignored. The text failed to mention the term "fossil fuels". Behind the words lay a colossal reliance on technologies to pull carbon dioxide out of the atmosphere at a scale that is unlikely to be feasible, economically, technically, or ecologically.

Incredibly, the 1.5°C (2.7°F) target survived the final round of negotiations, as did a commitment to following the best science. In fact, there was a specific clause requesting that the Intergovernmental Panel on Climate Change, an independent expert group consisting of thousands of scientists, produce a report on the 1.5°C (2.7°F) global warming target. The final text had somehow improved on the penultimate version. It is impossible to know if our intervention influenced proceedings, but it was certainly a moment of truth and transparency. Aviation and shipping remained under the carpet, and the text lost the critical clause of cutting emissions up to 95 per cent by 2050 to meet the 1.5°C (2.7°F) target. The media narrative changed, too. While praise was rightly heaped on the tireless negotiators and political leaders, this was tempered by the sober, scientific reality that the Paris Agreement left too much to interpretation and failed to provide a mechanism to drive urgent action now.

How fast must emissions fall?

In the weeks after the climate negotiations, we had some deep conversations about what it would take to meet the Paris Agreement. Very basically, the Paris Agreement says that the world aims to balance any greenhouse gas emissions in the second half of this century. What goes up must come down.

There is an obvious problem with the way this is framed: there is no sense of urgency and no pathway linking this current crop of politicians and business leaders, on their Earth watch, to this distant future, 30 to 80 years from now. It opens up a million pathways, and, deliberately or not, kind of indicates that an incremental change over a generation or two will be enough to solve the climate challenge. This suits the oil companies. Looking at one extreme scenario, you could argue that the world could wait until 2049 and hope that, by then, there would be a magical technology that would suck all our carbon out of the atmosphere in the nick of time. It is blatantly obvious that while there are millions of pathways, some are more plausible than others.

We started brainstorming a new scientific paper to more clearly articulate a pathway to reach the Paris Agreement. For reference, we used the agreement's goal of keeping the average temperature on Earth "well below 2°C (3.6°F) above pre-industrial levels and to pursue efforts to limit the temperature increase to 1.5°C (2.7°F)".

In reality, emissions must shrink to zero or close to it by about 2050 – not 2060 or 2070, and certainly not 2100. This would give us a reasonable chance of staying well below 2°C (3.6°F) global warming and a very slim chance of stabilizing at 1.5°C (2.7°F), or a little above. Leaving it too late means building a new industry, larger than the existing fossil fuel industry, to remove carbon dioxide from the atmosphere. It means planting trees on a scale that will compete with food production at the same time as the population is reaching 10 billion people. Most of the technologies that would be required for this either do not yet exist or have never been tried at scale. And this plan may not even work if the cascading tipping points are triggered and emissions keep rising from permafrost and forests. The "slow decline" solution is as realistic as a field of unicorns.

There is also a communications problem: anything with a deadline a generation away is not going to spark an emergency response immediately. It seems no one realized that in order to hit zero emissions by 2050, most action needs to happen right now – this decade. The only alternative is very likely economic collapse.

Rule of thumb

Serendipity is a powerful force for new thinking from cross-pollination of ideas. Around this time, we had already had a few conversations with Johan Falk, an executive at Intel, the Silicon Valley chip manufacturer. He was deeply concerned about climate inaction and felt that the world needed a new narrative to spur action. Falk encouraged us to think about climate action in terms of the Silicon Valley mindset of disruptive exponential growth that emerged in the 1960s. Intel founder Gordon Moore published a technical paper in 1965 showing that computing power doubles every 12 months, later revised to every 24 months – an exponential doubling trajectory. This paper had a disproportionate influence on the emerging tech sector: the simple doubling idea became the dominant guiding star for the entire industry. Thousands of companies and their supply chains were soon adopting it.

Incredibly, the tech sector has more or less stayed on this path for 50 years, disrupting everything and giving us the modern world. The doubling trajectory became known as Moore's Law, yet it is not a law of physics or a legal requirement. It is more like a rule of thumb. Complex systems are often governed by simple rules. Falk asked, "Is there a way to frame climate action that captures this powerful dynamic?" We said we'd have a think.

Exponentials do not just go up, doubling then doubling again; they also go down, halving and halving again – think of the half life of a radioactive particle. We had been focusing too long on the end of the emissions curves – way out beyond 2050 – instead of looking at the start. In the 1960s, NASA directed most of its efforts into getting astronauts off Earth in the Saturn V rockets, not the Moonwalk at the end. Instead of focusing on 2050 and beyond, it is obvious when you look at graphs of realistic emissions pathways that the lion's share of the action needs to happen now. After weeks of pouring over the data, it dawned on us that, very loosely, the world needs to halve emissions by about 2030 to have any hope of staying on the trajectory to stabilize the planet anywhere near 1.5°C (2.7°F) global warming by 2050. And then slash emissions in two again by 2040 and again by 2050. An exponential pathway. All the time, we need to build up the resilience of the rainforests and peatlands and start experimenting with ways to pull emissions out of the atmosphere. There is no doubt we will need these technologies.

This halving trajectory seems obvious now. It is widely talked about in the media and among NGOs (non-governmental organizations). But at the time, it was novel. It even seemed extreme: can the world really cut emissions by half in 10 years? In fact, at the time we were less interested in whether it was feasible. Rather, our aim was to communicate how unfathomably off course the world had become. International politics had come untethered from the physical reality of the planet.

You can spend all day debating the finer details. Should emissions be cut by 48 or 52 per cent? In reality, a rough rule of thumb is good enough. We dubbed the halving trajectory the "Carbon Law", after Moore's Law. Like Moore's Law, the focus is not on arriving at zero emissions; it is on the next step to take you there. This insight sparked our co-authors of the scientific paper, Hans Joachim Schellnhuber, Nebojsa Nakicenovic, Joeri Rogelj, and Malte Meinshausen, to think about what was needed politically and economically every decade in order to keep to this Carbon Law pathway.

Carbon Law pathway

The Carbon Law operates like a fractal. It applies to the whole world, but it also applies to a country, a city, a business, a family, or an individual. We all need to cut emissions by half. It also means that those who emit the most emissions have to cut the most. Rich countries, of course, have a responsibility to exit carbon faster in order to allow poorer countries to develop. But the Carbon Law offers two other critical insights. First, it makes climate action relevant for political elections and business cycles. The Carbon Law pulls thinking and political narratives away from distant goals over far-off horizons to focus on the here and now.

Second, some industry leaders often complain that they cannot reach zero emissions. It is impossible. So, they argue, more research

The future of carbon
Stabilizing global
temperature at around
1.5°C (2.7°F) means
emissions must peak
around 2020. At the same
time, we need to build
capacity to remove carbon
from the atmosphere.

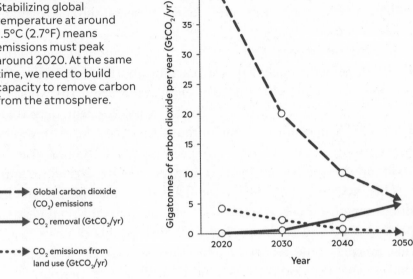

Global carbon dioxide
(CO_2) emissions

CO_2 removal (GtCO_2/yr)

CO_2 emissions from
land use (GtCO_2/yr)

and development are needed before action. Certainly, research and development are required for those industries that are hardest to transform: aviation, steel production, or cement production. But commercially attractive solutions to cut emissions by half exist in all industries now. We are already halfway to the Moon.

It was important for us that the Carbon Law goes beyond emissions reductions. Without this, we will fail in delivering the Paris climate targets. We have expanded the concept to include three more pillars.

First, farmers must turn agriculture from a vast emitter of greenhouse gases into a store. We will discuss this in Chapter 11. Second, we need to protect our remaining carbon sinks – the wetlands, forests, soils, and ocean. Third, we need to build more places to store carbon. This could be through reforestation or more high-tech solutions, for example. The scale must be manageable, though, and must not interfere with feeding 10 billion people or protecting biodiversity.

We published the paper "A roadmap for rapid decarbonization" in March 2017 in the journal *Science*. A month later, Paul Hawken published *Drawdown*, which describes 100 solutions to reach zero emissions. At first, some thought this was a competing idea, but in reality the two are perfectly complementary. *Drawdown* outlines the technical solutions, while our paper presents the pathway to implement the solutions and the policies to deliver them.

In 2018, the major assessment report requested by nations in the Paris Agreement concluded that reaching the 1.5°C (2.7°F) target was not only achievable but also essential to avoid the worst of climate change – 2°C (3.6°F) is far, far worse for humanity than 1.5°C (2.7F). It confirmed our findings that the best pathway to staying well below 2°C (3.6°F) and aiming for 1.5°C (2.7°F) is to slash emissions 50 per cent by 2030. The idea began to go mainstream.

This new framing may be influencing some of the public discussion on climate. If scientists are saying that emissions need to peak in 2020 and fall by half in 2030, then people can clearly see that precious little has been done thus far. We need to clarify, though, that the world will not end in 2030 if countries fail to cut their emissions by half. Societies will carry on. But they will increasingly struggle, as they become overwhelmed with cascading, compound, and cumulative impacts from climate change. If we continue to increase emissions at the rate of today and only start acting in 10 years' time, the economic path will be too steep. The economic and democratic options will be extremely slim. Without drastic emissions reductions, we face greater risks of crossing irreversible tipping points, as discussed earlier. We have two choices. Either we knowingly commit future generations to a dangerously destabilized planet, or we act on the scale needed to restabilize the planet to ensure that our descendants have what we have: the chance of a good life on a stable planet.

One unexpected upshot of all of this is that we joined forces with Falk and some big players in the technology industry, such as Ericsson – along with the WWF and the Future Earth project – to develop detailed "Exponential Roadmaps" that go deeper into the journey of halving emissions. We found that every single sector of the economy can cut their emissions by half relatively easily. Manufacturing: adopt a circular economy. Transport: focus on city transport and electric mobility. Food: reduce waste, eat healthily, farm sustainably. Buildings: insulate and use more efficiently. Energy: use wind and solar power, improve storage. Essentially, electrify everything. We have now produced two "Exponential

Roadmaps" (2018 and 2019) that present 36 viable solutions, which, if implemented, would cut emissions by half globally by 2030 – and many companies could go much faster. No more research is needed. While we follow this trajectory, though, we must also invest in the research to halve again beyond 2030.

The most striking finding from our work relates to energy. We all know that solar and wind have expanded rapidly. The common narrative states that they remain low in the energy mix – just 5 per cent – but not so long ago they accounted for only 0.5 per cent. Wind and solar mushroomed exponentially, doubling every three to four years. If this rate continues, then 50 per cent of the world's electricity will be supplied by these two sources alone by 2030. The critical issue is price. As long as prices keep falling, this trajectory will continue. Crucially, we have already crossed an economic tipping point: the price of wind and solar has crept below that of fossil fuels in many places. This is a game changer. Increasingly, it is cheaper to build a wind or solar plant than a fossil fuel plant. In parts of Europe, it is cheaper to build a solar plant than to keep existing coal plants running.

The stone age did not end because we ran out of stones. New technologies simply usurped them. This is the end of the fossil fuel era, even if oil and coal remain in the ground. Unlike giant power plants that require huge infrastructure and investment, wind and solar are small and flexible. This means innovation and roll-out are rapid. These are exponential technologies. The coal and oil industries innovate slowly. Soon they will be unable to compete.

Our aim with producing the Carbon Law was twofold. First, we wanted to provide a realistic pathway for the Paris Agreement. Second, we wanted to show the colossal scale of the economic shift required. The 1.5°C (2.7°F) target slips further from our grasp as each year goes by: literally, the remaining carbon budget shrinks more than 10 per cent each year if we continue to burn fossil fuels at the same rate as 2019. Without the right policies in place to accelerate action, such as a price on carbon, an end to fossil fuel subsidies, and tougher emissions standards, then we can forget about the global warming target altogether.

Very practically, emissions need to drop a massive 7 to 8 per cent every year. Emissions have collapsed rapidly before, but this usually accompanies economic shocks, for example during the 2008 financial crisis or the 2020 pandemic. We want to avoid this at all costs. Not least because it is economically destabilizing, which leads to political destabilization, and then we are back to square one. But is any country even close to the Carbon Law pace? Eighteen rich countries, including the United Kingdom, France, Germany, Ireland, and Sweden, have collectively reduced emissions 2 to 3 per cent a year for a decade or more. These countries have mainly achieved this by shifting to renewables, using greater energy efficiencies, and introducing a number of climate policies. Can nations double or even triple the rate of change? Most of the energy transition to date has largely focused on electrical power and wind and solar. If countries adopt, in parallel, measures to transform transport, building, cities, food systems, and manufacturing, then much larger changes are undoubtedly possible.

Energy is the first of the six Earthshot system transformations. When we left the COP 21 conference centre back in 2015, no major economy had committed to reach net zero emissions by 2050. Now, the United Kingdom, France, New Zealand, and others have enshrined this target in law. Furthermore, the European Union is the first continent to adopt a net zero target for 2050. Sweden has gone further and is aiming to reach net zero by 2045. Finland plans to reach it a decade earlier, in 2035. Norway has its sights on 2030. What's more, in 2020, we received the most surprising news imaginable. The Chinese president, Xi Jinping, announced to the United Nations that China aims to become "carbon neutral" before 2060. China. The world's largest emitter of carbon dioxide. This is a game changer.

These are all monumental steps. We are locking in a future with zero emissions. This is the future we want.

FEEDING 10 BILLION PEOPLE WITHIN PLANETARY BOUNDARIES

Our world is falling apart quietly. Human civilization has reduced the plant, a four-hundred-million-year-old life form, into three things: food, medicine, and wood. In our relentless and ever-intensifying obsession with obtaining a higher volume, potency, and variety of these three things, we have devastated plant ecology to an extent that millions of years of natural disaster could not.

HOPE JAHREN, *LAB GIRL*, 2016

When we published the "planetary health diet" back in 2019, Brian Kahn, a US journalist writing for the science and technology media website Gizmodo, decided to follow it for a month. Towards the end, he reported, "My 30 days is almost up, and honestly, I think I'll keep the diet up for the most part. For me, it wasn't that hard, and I really value the health benefits side of eating more whole foods in general."

But it was not all plain sailing. Kahn caved in to a doughnut and pizza craving in a moment of stress one evening. We've all been there many times. No judgement. This forced him to reflect that the diet is as much about personal health as it is about the health of the planet. And this is precisely the purpose of the diet: to provide general guidelines for a healthy life on a healthy planet as we grow to 10 billion people.

Food determines our future on Earth. Fail on food, and we fail both people and planet. This is why food is the second of our system transformations, and alone it will ultimately determine whether we successfully meet the Paris Agreement goal of keeping global warming below 2°C (3.6°F). Food also plays a key role in whether we are able to reach the United Nations Sustainable Development Goals. Remember, all 17 goals, from poverty and hunger to the ocean and land, need to be achieved by 2030. We are talking of nothing less than a planetary food emergency.

This may sound dramatic. Is food really failing us that badly? Does food alone truly hold the key to success? Well, the short answer is yes. The battle for climate change is no longer being fought over the global energy system. Decarbonization is on its way and is relatively simple compared with food. No, the final battle over whether we successfully deliver on the Paris climate targets rests on whether we are able to transform the global food system.

Both transitions are needed to succeed; that is undeniable. But right now, the energy transformation – despite all our concerns, the frustratingly slow progress, and the lack of urgency – is far ahead of food. The food transition is lagging behind in terms of policy, economy, awareness, and technology. In fact, we would argue that the topic of sustainable food is today, in 2021, where the energy agenda was 30 years ago. We do not, as yet, have the policy tools, political debate, and solutions to change course. Indeed, many countries are moving towards a Western junk food culture that threatens not only people's health but also the stability of the planet.

Things are beginning to change, though. It feels as if we are witnessing a sudden awakening. COVID-19 provided a devastating reminder that our food system, in this case the trade in wild animals and the proximity of farms to other ecosystems, is both vulnerable to shock and a driver of instability. Today, in science, policy, business, and the media, there is a greater focus on food, health, and the planet, perhaps greater than ever before. But it is late in the day, and so far, we see many scattered islands of new ideas in a sea of business-as-usual behaviour.

A broken food system

As an Earth scientist, Johan has spent many years assessing the role of food in the sustainability of our ecosystems and, increasingly, how environmental impacts from food production combine to affect the entire planet. It has been quite a staggering journey.

The way that we produce food in the world is the single largest reason that we have transgressed planetary boundaries. It is the single largest threat to the stability of the planet and our life support systems, from fresh water, pollinators, and soil health, to rainfall generation, and quality of air and water. Food production is putting our future at risk.

The agricultural enterprise is so large it constitutes a global force of change on Earth. A great amount of research and development is going on to find ways to farm sustainably. But the focus has largely been on reducing the environmental impacts of local farming practices. This has led to improvements in water use efficiency,[1] and solutions to reduce leakage of nitrate and phosphate into groundwater and rivers. Even though we have improved water efficiencies and reduced environmental impacts, we have focused on one thing only: increasing production.

The impacts of our food system are amplified because of the sheer scale of global agriculture and the way in which the food system eats into all corners of our societies. Take nitrogen, for example. The European Union has good environmental regulations on all forms of reactive nitrogen flows, whether it be in the air or in fresh water. Agriculture is responsible for up to 60 per cent of European nitrogen pollution into groundwater and rivers, mainly through the use of manure and fertilizers. Although the regulations are important, they do not consider the wider system. A large part of reactive nitrogen is often imported from other continents as feed

---

1 We now get more "crop per drop" by using techniques such as mulching, drip irrigation, and nutrient management.

for animals and as fertilizers. For example, soya from Brazil, high in nitrogen, travels halfway around the world before being eaten by Swedish cows and then leaked onto Swedish soil and into rivers and out to the Baltic Sea.

Furthermore, most of the nutrients applied to farmers' fields are exported to and concentrated in cities. The crops that are grown with the imported nutrients are harvested for food. Consumers eat the food. Most consumers live in cities. The food ends up either as food waste or as human excreta, channelled through waste water treatment systems with varying degrees of efficiency.[2] Too many nutrients cause pollution and eutrophication (excess algal growth and a fall in oxygen in waters) in coastal zones and lakes furthest downstream. This is occurring all over the world. The modern fertilizer industry is so large that it injects more reactive nitrogen into the biosphere than the entire natural nitrogen cycle. This linear system is undermining our world. We need a wider planetary boundaries approach to make agriculture sustainable.

Food as the No. 1 killer

Food is destroying our health and shortening our lives. It is the single largest killer, responsible for more deaths than smoking, AIDS, tuberculosis, and terrorism combined. Three independent research studies in 2019 estimated that 11 million people in the world die prematurely because of unhealthy food. The fastest-growing killers are obesity and diabetes. In Asia, 2.4 million people die from diabetes each year. Even in developing countries, obesity rates are on a par with those for undernutrition. In Indonesia, for example, there are more overweight than underweight people. Obesity-related diseases were long believed to be exclusive to wealthy countries, but now more than 70 per cent of the world's 2 billion overweight and obese individuals live in low- or middle-income countries. Faced with increasing disability, mortality, and health care costs, as well as lower productivity, all countries are finding obesity a growing concern, regardless of income level.

---

2 The most modern systems generally lose up to 30 per cent of nitrogen and phosphorus.

In response to the realization that food is threatening not only our planet but also people's health, Gunhild Stordalen created the EAT Forum in 2013, with scientific guidance from Johan. The initiative is essentially a World Economic Forum for food, held in Stockholm. At the first event in 2014, we established that we urgently need a global scientific assessment of the state of knowledge on health and sustainable food. Major intergovernmental scientific assessments on climate and biodiversity take years to gain traction, then each report takes six years to compile. If we take this route, it may be 2030 before such a food assessment sees the light of day. We simply do not have time. Instead, in discussions with Walt Willett,[3] we agreed that we need a faster, more agile assessment that focuses on solutions. While we were debating the massive knowledge gaps on health and sustainability for food, Richard Horton, editor-in-chief of the leading medical science journal *The Lancet*, was also in the room. He encouraged us to write to *The Lancet* and propose a global integrated science assessment. This became the EAT-*Lancet* Commission, tasked to synthesize our knowledge on healthy diets from sustainable food systems. It showed that we will fail with the Paris Agreement if we fail to transform food from the single largest source of greenhouse gas emissions to a major carbon sink. The commission set out to define two things for the first time: the safe boundaries for the health of both people and planet, and a universal diet, called the planetary health diet – in short, what science can tell us about the options we have left if we want to live healthy lives on a healthy planet.

Food alone threatens the stability and resilience of our planet

Food is by far the largest consumer of fresh water. A staggering 70 per cent of all the withdrawals of fresh water from rivers, lakes, and groundwater is used to produce food. The average person needs between 50 and 150 litres (13 and 40 gallons) of water per day, to shower, wash clothes and dishes, and flush the

---

[3] Professor of epidemiology and nutrition at Harvard University and co-chair of the EAT-*Lancet* Commission.

toilet, but our food requires some 3,000 to 4,000 litres (793 to 1,057 gallons) per person per day, to grow crops and fodder. Rivers like the Colorado and the Limpopo, as well as the Aral Sea, are running dry, due to over-irrigation of food crops. Warmer global temperatures, caused to a significant extent by agriculture when we burn fossil fuels, cut down forests, and degrade land, trigger more serious droughts, floods, and heatwaves. These extreme events further increase water scarcity. Vulnerable communities are particularly at risk, especially small-scale, rain-fed farmers in savanna regions.[4] Food insecurity, exacerbated by climate change, is a breeding ground for geopolitical conflict. The Arab Spring, the conflict in Syria, and the civil war in Sudan are all examples of social unrest, and ultimately political collapses, linked to dramatic impacts on food production systems that were caused by droughts made worse by climate change.

We are currently in the sixth mass extinction of species on Earth, with one in eight species at risk. The way we produce food on land and catch fish in the ocean is the main driver of this extinction. We have already transformed 50 per cent of Earth's habitable land beyond glaciers and deserts for different forms of agriculture. This is the prime reason why we are losing species and healthy ecosystems. Although agriculture is causing this, it is also in the firing line. When we lose trees, pollinators, earthworms, and wild predators, our landscapes lose their resilience. When this happens, they can no longer provide the functions needed for food production.

Food is the main reason that we are transgressing the most critical planetary boundaries: land, biodiversity, climate, and nutrients. It is threatening the stability and resilience of our planet. Without serious attention, we face multiple crises. Lifting 1 billion people out of poverty and hunger and feeding another 2 to 3 billion new citizens by 2050 will require some 50 per cent more food. If we are knee-deep in problems today, just imagine where we will be in 30 years' time.

---

4 Savanna regions cover 20 per cent of the world's surface and host a large portion of the world's population.

Food production footprint
Land use per 100g (3.5 oz) of protein. Red
meat production requires about 100 times
more land per gram of protein than cereals.

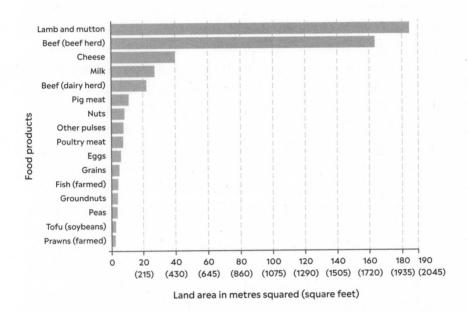

Land area in metres squared (square feet)

The planetary health diet

The aforementioned EAT-*Lancet* Commission was established to
find solutions for this planetary drama. It was the first attempt to
scientifically define quantitative targets for a healthy and
sustainable diet. If we want to maximize our chances of good
health and a healthy planet, there are safe operating spaces for all
people, geographies, and cultures around the world. Through the
EAT-*Lancet* Commission, we defined these safe operating spaces
– the multitude of diets that are most sustainable – for the first
time, using real numbers based on science. As a result, in 2019, the
EAT-*Lancet* Commission released the "planetary health diet", to
maximize our chances of good health on a stable planet.

For human health, a flexitarian diet is recommended, and the
planetary health diet advises eating five servings of meat or fish
each week.[5] We should be eating a lot more nuts and legumes, fruits,

---

5 One from red meat and two each from chicken and fish.

vegetables, and wholegrain foods. At the same time, we need to reduce our intake of saturated fat (found in fatty meats, pastries, cheese, and so on), dairy products, starchy vegetables like potatoes and cassava, and salt.

In some ways, the planetary health diet is similar to the traditional Mediterranean diet: rich in fresh fruit and vegetables. Some people raised concerns that the planetary health diet suggests everyone around the world should eat the same thing: a "one size fits all" approach. Nothing could be further from the truth. The planetary health diet merely sets the scientific boundaries for a healthy diet for people and planet. How you follow it, and to what extent, is your own decision. Anything else would fundamentally disrespect human dignity. The authors felt very strongly that everyone has a right to know what the science tells us, while recognizing that food is deeply rooted in local culture and traditions. We deserve to know what is killing us and what can prolong our lives. A myriad of different food cultures, from Asian marine diets to African savanna diets, can easily be accommodated in the planetary health diet.

The real excitement with the planetary health diet, though, comes from integrating the state of our planet with the planetary boundaries for food. If we eat according to the planetary health diet, we significantly prolong our lives and improve the stability of the planet at the same time.

Can we feed 10 billion people within planetary boundaries?

The world population is heading towards 10 billion people by 2050. By this time, land degradation and climate change together are predicted to reduce crop yields by an average of 10 per cent globally and up to 50 per cent in certain regions. Can we really feed this many people and eradicate hunger without breaking the planet? If we follow the planetary health diet, the pressures on the planet would drop precipitously. However, eating healthy food alone would not bring us back within a safe operating space on Earth. The EAT-*Lancet* assessment shows that we need to take two key actions in addition to eating healthy food. First, we need to reduce food waste at every stage: this accounts for about 6 per cent of all greenhouse gases.[6] Second, we need a global transition towards farming practices that capture rather than lose carbon, that circulate nutrients rather than pollute, and that save water rather than waste it. We need sustainable agriculture. And the good news is that an agricultural revolution is under way. According to a recent study by Jules Pretty at the University of Essex and colleagues, 29 per cent of farmers worldwide are already practising some form of sustainable agriculture. We are on the right track in some places, but we need to accelerate.

Big picture solutions

Science can guide us in speeding up the transformation. For the energy transition, we introduced the Carbon Law as a guiding star. For food, we need a similar principle. We suggest that the guiding star for the food system transformation is the number zero. Zero is powerful. The state of ecosystems, both on land and in the ocean, is so dire that we need to adopt a zero target for nature. We need zero loss of nature from now – 2021 – onwards. All future losses of forests, ecosystems, and species must be restored and regenerated. We need to safeguard the remaining carbon sinks, ecological habitats, and rainfall-generating systems on Earth. As we have already transformed 50 per cent of Earth's land-based ecosystems to agriculture, cities, and roads, our task is to feed future populations from our current farmland. We must now ensure zero expansions of new agricultural land. For nature, we will unfortunately not be able to halt all loss from now onwards. From now on, though, all nature that we lose must be restored by 2030. By 2050, we need to be regenerating more nature than we had in 2020.

What does a zero target mean for future food production? Feeding humanity from our current cropland while at the same time storing more carbon on land can only be achieved through sustainable intensification. This implies improving yields while meeting all planetary boundaries. A series of different practices can be used, thus paving the way for nothing less than a new "green-green" revolution.

---

6 Every year, we waste approximately 30 per cent of the food we produce.

This involves:

- Landscape planning. Farming practices can no longer focus on monocultures; they must consider ecosystem services, including water management, soil health, protecting pollinators, storing carbon, encouraging diversity, and balancing livestock and crop systems.
- Circular production systems, for carbon, nutrients, and chemicals.
- Decarbonizing agricultural energy use, for all energy inputs.
- Conservation agriculture for carbon storages. This means abandoning ploughing and investing in minimum tillage and mulch farming.

Ultimately, the foundation of the future of farming is built on regeneration and recirculation – as if we live on a spaceship where we treat the food system as part of the life support system. Building resilient and productive sustainable farming systems requires planning at the landscape and watershed scale.[7] We need to safeguard all ecological functions, such as water flows, pollination, and soil health. We need corridors for wildlife; we need wetlands and natural forest systems for moisture feedback and as nutrient sinks. We need carbon sinks and a diversity of grass, trees, and animals in the food web.

Many farmers already practise recycling nutrients, biomass, and water. They are integrating livestock with cereal production, producing their own bioenergy for traction and mulch, and providing their own animal protein feed, which is circulated back onto their own farmland. We need to expand these practices.

For centuries, we have been ploughing fields with ever more powerful machinery. But it turns out that this technique actually reduces soil quality. One of the most exciting transitions of the food system is replacing ploughing with different forms of minimum tillage. The idea is to avoid turning the soil. When the soil is turned, the microbes and carbon in the topsoil are exposed to the Sun and wind, which causes the soil to lose organic matter and kills the good

---

7 It does not matter if you are a small-scale, rain-fed maize farm in Burkina Faso or a large mechanized farm in Russia.

microorganisms living in the soil. Conservation tillage copies nature by building up biologically rich soils. Turning the soil, particularly in tropical regions, leads to loss of carbon. Stopping ploughing allows soils to build up organic matter and soil biology. In addition, ploughing uses a lot of energy, either as diesel for tractors or fodder for oxen.

Different techniques have been developed to open up soils only where seeds are planted. Farmers can use various types of rippers or sub-soilers to cut a narrow planting line, without touching the rest of the soil. This approach allows them to place manure and fertilizers exactly at the spot where seeds are planted, without broadcasting nutrients and fertilizing both crops and weeds. Small-scale farmers in developing countries generally find conservation tillage very attractive, given their limited resources. An example of a traditional conservation tillage practice in West Africa are the zai-pits. Farmers dig shallow pits instead of ploughing whole areas. When it is time for planting, these become planting pits. This technique concentrates all the nutrients and water in the optimum spot, thereby minimizing leakage. It is a form of manual precision farming, just like modern farmers today use advanced technology for the precise application of fertilizers.

From a planetary boundaries perspective, an interesting positive side-effect of conservation tillage is that it helps transform farms to store carbon. These techniques can convert farms from major carbon emitters to essential carbon sinks, while increasing soil fertility. This is a win-win situation for both farmer and planet.

Undoubtedly, the food system challenges we are facing are huge. We have to explore integrated system solutions that combine water productivity, soil health, nutrient recycling, crop rotations, and watershed design with advancements in biotechnology. To increase the volume of food without expanding agricultural area in a sustainable way requires not only new practices, but also new crops. Imagine wheat that does not need to be planted each year? Perennial cereal crops are a step closer to this reality. If these crops are successful, it will reduce tillage and allow crops to develop deep root systems, thus storing more carbon and building more resilience to water stress. We can see this trend in Kenya, for example, with the expansion of perennial legumes like pigeon pea. This crop grows over several years, provides a very tasty and healthy food, and has deep, powerful roots.

There are two planetary boundaries on which food has a major impact that we have so far been unable to quantify: the aliens of Act II, novel entities and aerosols. Thanks to herbicide and pesticide use, agriculture causes persistent organic pollutants and endocrine disruptors (chemicals that alter hormone levels in animals, including humans) to build up in the biosphere. Any transition to sustainable food systems must minimize the impacts of these novel entities. We have a lot to learn here from ecological agriculture. Many modern, large farms in the United States, Europe, and Asia are already successfully producing food without pesticides.

Across many parts of the world, farmers annually burn the remains of the last crop to prepare the land for the next one. At an Anthropocene scale, so-called slash and burn is immense. Vast clouds of smoke billow over large regions, engulfing megacities such as Delhi, where the chemicals mix with Delhi's own pollution issues and bring a deadly cocktail deep into the lungs of millions of people. Aerosol pollution (small particles in the atmosphere) from agriculture is a major threat to human health, regional climate stability, and weather systems. Agriculture is also responsible for sulphate and nitrate emissions from diesel burning and black carbon emissions from biomass burning. Small-scale slash and burn farming must be abandoned – and there are plenty of alternatives that improve yields – if we are to flip agriculture from a carbon source to a carbon sink.

The next decade: what needs to happen, what is possible

This next decade, from 2020 to 2030, must be a turning point. Just as we have to cut global emissions by half in 10 years, we need to commit to 100 per cent sustainable agriculture and food systems around the world in the same timeframe.

This must be the decade in which we invest all our efforts in producing sustainable food from existing farmland. The 10,000-year-long era of agricultural expansion has now come to an end. We draw this conclusion from our Earth system analysis of the rising risks of destabilizing the planet. Already we have transformed half of Earth's surface. We must keep the other half intact, in line with Edward O. Wilson's "half-Earth" philosophy. We therefore need to take a planetary stewardship approach to food transformation. Every hectare of cultivated land must move from being a carbon emitter to a carbon store, and needs to host a better diversity of crops, wild animals, and insects. Every hectare of land needs to have an inbuilt resilience to deal with unavoidable shocks, such as droughts, floods, cold spells, and heatwaves. Sustainable agriculture must, by the end of this decade, be the new normal.

Technologies, policies, and regulations will play a critical role in this transformation. We require advancements in resilient crops with better drought resistance and carbon sequestration potential. We talk a lot about a global price on carbon. It is direly, so direly, needed. Forty-six countries have adopted a price on carbon. Yet not one applies it to food, despite the carbon gushing from food production into the air. We need to start accounting for the negative impacts of food systems around the world, not just locally. This involves more than a price on carbon; we must also consider a price on nitrogen, phosphorus, and water. For this, we require international agreements that regulate the protection of the remaining natural ecosystems on Earth. We need to bring further loss of ecosystem functions to zero. This is critical.

Furthermore, we must restore and regenerate the destroyed land. Worldwide, land degradation is shocking. Soils in croplands have lost up to 60 per cent of their organic carbon and are reaching a point in places where neither food nor biodiversity are viable.

Competing demands for land from food, renewable energy, and nature-based carbon sinks are inevitably causing rising pressure. This is a massive challenge. There is no silver bullet, though. We need to pull out all the stops. While nature-based solutions and tree planting for carbon sinks and biofuels are part of the "solution space", they are not able, under any circumstances, to replace the need to abandon fossil fuels, or even to reduce the pace at which the world needs to do so.

The message is clear. To stand a chance of deviating away from the climate disaster, and to maintain the ecological capacity on Earth to feed humanity, we need to follow both the Carbon Law for energy and the zero law for nature.

INEQUALITY IS DESTABILIZING EARTH

Society can't function without shared prosperity.

JOSEPH STIGLITZ

WINNER OF THE NOBEL MEMORIAL PRIZE IN ECONOMIC SCIENCES, 2001

Let's begin with a small tribute to a uniquely wonderful Swedish academic: Hans Rosling, who sadly died in 2017. Rosling was a medical doctor and researcher. But more than that, he was a force of nature. He tirelessly promoted a fact-based world view to audiences at the World Economic Forum and the World Bank, as well as on TED stages, discussing health, economic development, and poverty. His presentations were riotous and revealed deep truths about the audience's own perceptions of the world.

Rosling often began a talk by grilling the audience with the following question: "In the last 20 years, the proportion of the world population living in extreme poverty has ...

A: almost doubled

B: remained more or less the same

C: almost halved?"

The correct answer is C, but polls show that less than 10 per cent of people are aware of this remarkable achievement. The number of people living in extreme poverty is reducing exponentially. This is the biggest success story of the past century. In 2020, just 8 per cent of the global population lived in extreme poverty. Over the past 25 years, more than 1 billion people were lifted from extreme poverty.[8] This is impressive, but extreme inequality is another matter entirely.

An online video of two capuchin monkeys went viral a few years ago. They are sitting next to each other in two separate cages and have been trained to do a small task – hand a stone to a researcher – in exchange for a treat. At the start, the researcher hands them slices of cucumber in return for the stones, and they seem content to work away for the reward. Then, one is given a grape instead of a piece of cucumber. Grapes are a far superior treat for monkeys. The reaction of the second monkey is amusing. It watches suspiciously and checks its stones for deficiencies. Becoming agitated, it grabs a slice of cucumber and hurls it at the researcher. It refuses to perform the same task for a substandard reward.

If grapes were dollars, Amazon employees on a minimal wage would earn 15 grapes an hour while company founder Jeff Bezos would rake in 4.5 million. In our digital world, our social media feeds remind us daily of the size of other people's grape mountains. This inequality and unfairness is politically destabilizing.

In this chapter, we have three insights on inequality – the third system transformation – that we want to share.

The first insight is that reducing inequality may be the single most important economic and political solution for planetary stewardship. But inequality is not only about poverty eradication and "developing economies" catching up with so-called "developed economies". It is also about a fairer distribution of wealth in the world at the level of families. Greater equality creates greater solidarity, making it easier for societies to come together around shared goals. The Earthshot is the ultimate shared goal.

The second insight is that making any progress to reduce inequality will be challenging in the next decade. This is in no small part due to how gross inequality has skewed politics hard right in many countries. Demagogues or even authoritarian leaders have risen from nowhere to gain power. We prefer the term "demagogue" to "populist" because populist leaders, in theory, aim to support the interests of the working people over elites. There is no evidence for this among the current crop of prominent so-called populist leaders.

---

8 Incredibly, however, extreme poverty in the United States, which has the world's richest economy, has increased. Between 2010 and 2016, some 1 million people are estimated to have descended into this economic bracket.

These leaders claim to want to "drain the swamp" and give more power to ordinary people, but their policies tell a different story and are likely to increase inequality.

The third insight is that for too long the media have treated economic inequality and the environment as two separate problems.[9] But this is a false split. Degraded environments reinforce inequality, and solutions to reduce economic inequality can also enhance planetary stewardship. Such actions will improve cooperation, build trust, moderate consumption, help collective decision-making, and spur innovation. And the best part? A stable, resilient planet will make it easier for the poorest in society to thrive and prosper.

Shocking inequality

It is now obvious that rapid economic development and the emergence of a global middle class have come at a high social and environmental price. We are living through a classic Catch-22 situation. Globally, economic growth has brought people out of poverty. But this has created profound inequalities within societies and driven environmental destruction. The solutions to the latter risk austerity measures that make the poorest in society poorer. In 2017, France's president, Emmanuel Macron, knew he needed to rein in emissions and slapped a tax on fuel. Diesel prices shot up 23 per cent in a year, hitting the poorest hardest. This move pushed people into fuel poverty and provoked anger, distrust, and discontent. In October 2018, the "Yellow Vests", or "Gilets Jaunes", took to the streets of Paris and other cities in France in protest. Prolonged mass demonstrations attracted millions of people. Eventually, Macron caved in and reversed his decision.

Badly designed policies to reduce emissions can backfire, driving greater inequality and deep resentment of ruling elites. But we do, in fact, know how to design economic policies without hitting those with low incomes. Had Macron given Stefan Löfven, prime minister of Sweden, a short call before introducing the carbon tax on fuel, things would probably never have erupted as they did. Löfven

---

9 Sure, economists who bother to look at the environment do link these two problems, but they often view environmental deterioration as a trade-off with economic development.

NOVEL ENTITIES

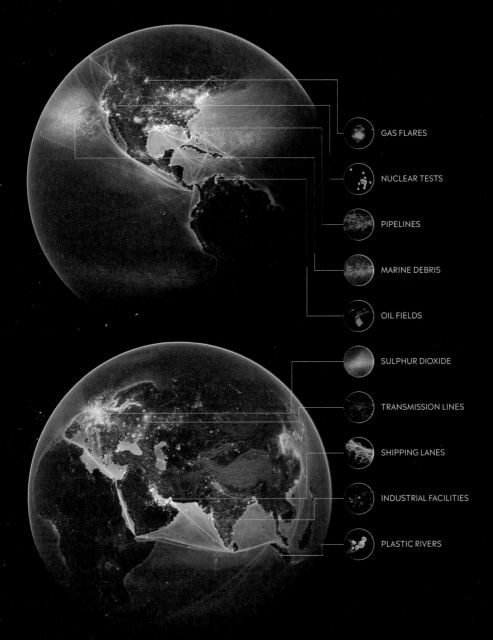

GAS FLARES

NUCLEAR TESTS

PIPELINES

MARINE DEBRIS

OIL FIELDS

SULPHUR DIOXIDE

TRANSMISSION LINES

SHIPPING LANES

INDUSTRIAL FACILITIES

PLASTIC RIVERS

"Novel entities", one of the nine planetary boundaries, refers to things like plastics and other chemical waste in the environment, but also to nuclear waste, genetically modified organisms, nanomaterials, and even artificial intelligence. Novel entities are substances created or modified by humans that did not exist in the biosphere previously (or certainly not in their new states). There are more than 100,000 novel entities in circulation. The sheer number and complexity of their interactions mean that this planetary boundary is yet to be quantified. Many novel entities may be benign, but some are most definitely not.

THE NETWORK EFFECT

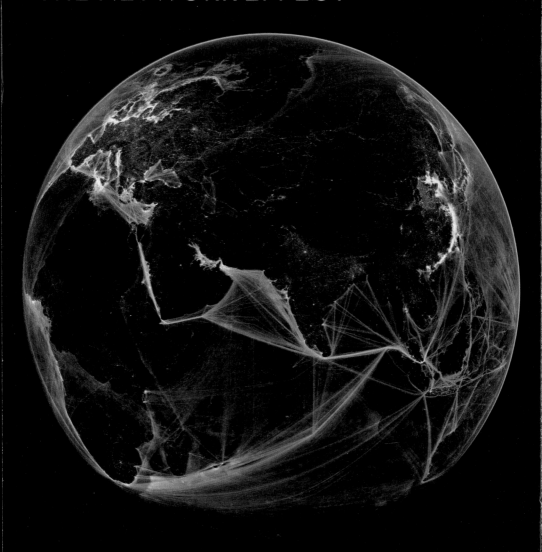

The story of humanity is one of travel, migration, trade, and connection.

This image shows humanity's industrial footprint on Earth in the Anthropocene: paved and unpaved roads, railways, electricity transmission lines, pipelines, shipping lanes, fishing boats, and undersea cables, all connecting human settlements (the lights of towns and cities). Our most densely populated areas are along coasts, on deltas, and close to rivers. Trade routes have long connected civilizations; now the interconnectivity is so deep and all-encompassing that we can say there is just one single civilization on Earth.

We are a big world on a small planet.

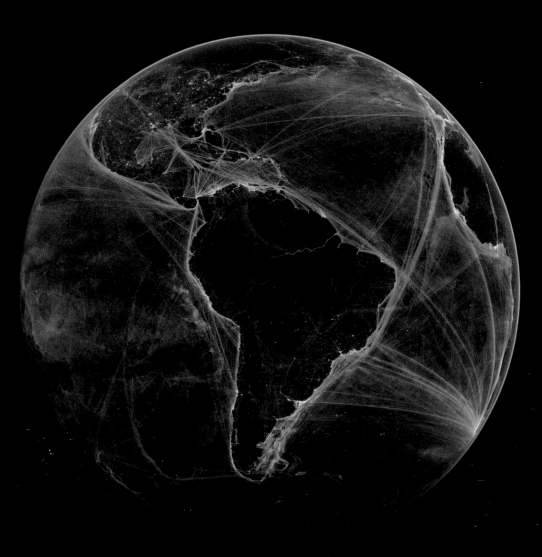

A major rupture shook the world 500 years ago when European settlers arrived in the Americas. This event connected previously unlinked human societies, which, in turn, led to twin ecological and cultural shocks that still reverberate today. This tightly interconnected world we have built brings benefits in terms of knowledge, culture, and prosperity. Ideas can spread rapidly, but so, too, can viruses, disease, and economic shocks. As the economist Jeffrey Sachs notes in his book *The Ages of Globalization* (2020), it took 16 years for the bubonic plague to spread from China to Italy in the 14th century. COVID-19 spread in a matter of days, carried by plane directly from Wuhan to Rome. Within four months, half of the global population of 7.8 billion people was in some kind of lockdown. This interconnectivity brings a degree of resilience – if crops fail in one place, we can ship them from elsewhere – but it also brings a new risk that we need to learn to manage: network fragility.

C3

A PLANET
TRANSFORMED

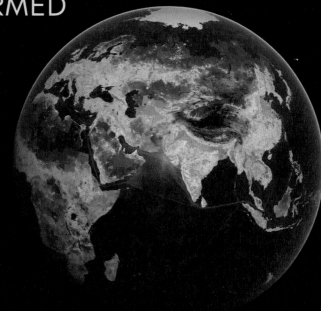

GLOBAL HUMAN
MODIFICATION

— Fully modified

-- No modification

Humans use half of all habitable land on
Earth for agriculture. One thousand years
ago, we farmed less than 4 per cent.

The global human modification key
(above) aggregates population density,
built-up areas, croplands, livestocks,
transportation, mining, energy production,
and electrical infrastructures.

here on in, those at the bottom and middle saw wages stagnate or decline, whereas those at the top saw incomes go through the roof. While many of us are better off than our grandparents in terms of health and longevity, it is the scale and blatancy of income inequality that has seeded the resentment and political instability that we see around the world today: social media feeds bombard us with images of rich people leading carefree, extravagant lifestyles. If you are not jetting to Bali or New York every weekend, you are a failure.

Inequality in the Anthropocene

Life will be made more difficult for many due to rising inequalities in the Anthropocene. Poorer countries are already vulnerable and will increasingly be hit hardest by the social instabilities and extreme events brought by the climate crisis. Food shortages are a key amplifier of inequality. Drought and flooding contributed to severe food shortages in southern and eastern Africa in 2019 and 2020, affecting 45 million people.

Inequality also impacts the lives of the poor in less obvious ways. As the ocean warms, the fish populations will move out of the tropics in search of cooler waters. Many poor people depend on fish for their livelihoods and food. As most of the world's poor live in the tropics, the decrease in fish populations in these regions will hit them particularly hard. In a similar vein, crop yields may improve in countries in the far north, as winters become less harsh, but land in the tropics will become harder to farm as temperatures rise. Rangelands in West Africa are projected to decline by up to 46 per cent, affecting 180 million farmers. But it gets worse. A 2020 assessment shows that for each 1°C (1.8°F) global warming, an additional 1 billion people will be living in places with intolerable heat levels – beyond the human climate niche. On our current trajectory, in only 80 years, this will put 3 billion people at risk of enduring extreme environments, for at least part of the year. This will predominantly hit arid and semi-arid hot tropical regions hardest: that is, regions hosting the world's poorest societies.

Climate change has already made it more difficult for poor economies to catch up. Economists first predicted this impact some years ago. A remarkable recent analysis indicates this has become a reality. Two academics at Stanford University, Noah Diffenbaugh and Marshall Burke, showed that poorer countries found it significantly harder to catch up economically with richer ones over the past 50 years because of climate change. An earlier analysis by Burke and other colleagues noted that economic productivity peaks when the average temperature is 13°C (55°F). The average annual temperature for wealthy countries is below 13°C (55°F), so they will benefit from a little global warming, whereas poorer countries tend to be warmer than 13°C (55°F) already and can expect productivity to decline. You can see where this is going: Red Queen economics.[11] Poorer countries already have to work harder to achieve the same level of economic productivity; in future, it will get worse.

Of course, wealthy countries are not immune to extreme weather. When disasters strike, though, it is the low-income neighbourhoods that suffer most: in the United States, nearly half a million government-subsidized homes are built on flood plains. And it is not just climate. Across Europe, poor neighbourhoods have higher air pollution than richer areas. In the United States, too, poor and minority communities are more likely to breathe in more dangerous air. African Americans, Latinos, and Asian Americans are exposed to 66 per cent more air pollution from vehicles than white people, a striking example of environmental inequality and racism.

Most research on inequality and climate focuses on the poor. Much less attention is paid to the rich and their behaviour. As people become wealthier, their ecological footprint expands. A team at Potsdam Institute for Climate Impact Research, led by Ilona Otto, estimated that the super-rich have an annual footprint equivalent to about 65 tonnes (72 tons) of carbon dioxide per person. This is more than 10 times the global average. Otto calculated that the emissions of the wealthiest 0.5 per cent of the global population are greater than the world's poorest 50 per cent when it comes to lifestyles. Given that Otto struggled to find many super-rich willing to talk to her about their carbon footprints, it would not be surprising if this is a gross underestimate.[12] Ironically, the super-rich can easily

---

11 In Lewis Carroll's *Through the Looking-Glass* (1871), the Red Queen tells Alice, "It takes all the running you can do, to keep in the same place."
12 One interviewee was not super-rich himself; he was a pilot of a private jet for the super-rich.

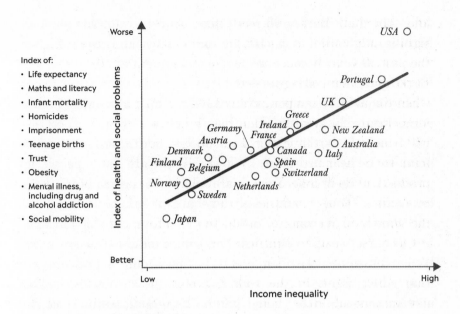

Health and social problems
People living in wealthy countries with high inequality suffer from more health and social problems than those living in more equal wealthy countries.

afford to buy solar panels and energy-efficient solutions to reduce their footprints (although flying accounts for more than half of their carbon emissions).

Equality and planetary stewardship

Can greater income equality support planetary stewardship? Yes. And we believe this is one of the most important untold stories in the Anthropocene. In 2009, two epidemiologists from the north of England, Kate Pickett and Richard Wilkinson, crunched the data on inequality across wealthy societies in their remarkable book *The Spirit Level: Why More Equal Societies Almost Always Do Better*.[13] What they found is astounding. People living in more equal societies, such as Sweden or Denmark, have better health, more social

---

13 In the United States, the book had a different subtitle: *Why Greater Equality Makes Societies Stronger.*

cohesion, more trust, and less crime. And in countries such as Norway, Finland, and Japan, life expectancy tends to be higher than in very unequal countries such as the United States, the United Kingdom, or Singapore.

More equal societies tend to do better in maths, literacy, and other educational achievements. They have lower obesity rates. There are fewer teenage births. Drug and alcohol addictions are not as prevalent. The murder rate is lower. There are fewer people in prison. And there is greater social mobility: if you are born into a low-income family in Sweden, there are more opportunities for education and training to allow you to move into high-income brackets than, say, in Brazil or the United States. You are more likely to live the American Dream in Stockholm or Copenhagen than in Los Angeles or New York. Pickett and Wilkinson also looked at inequality within the United States. They found the same pattern. States with greater income inequality did worse on everything, from opioid use to obesity.

The most remarkable conclusion, though, is that the richest people in more equal societies do better and live happier, more fulfilling lives than the richest people in unequal countries. Think about this for a minute. If Warren Buffett or Michael Bloomberg lived in Sweden, chances are they would be happier and healthier than they are living in the United States.

So, what gives? It is all to do with status. People's choice of clothes, films, and food, where we go on holidays, how we travel, where we live, in fact everything we consume says something about our status and self-esteem. We are extremely sensitive to status. Low social status is linked to poor health, drug use, depression, and decreased longevity. As Pickett and Wilkinson note, greater inequality heightens social anxieties by increasing the importance of social status. Marketing, branding, and advertising industries use our anxieties and insecurities to make us buy things that we don't need to impress people that we don't care about. It does not matter if you are rich or poor in this system; you will still feel insecure and unhappy.

So, how does this link to planetary stewardship and finding a safe operating space for humanity?

Inequality is a big driver of materialism. We show our status in society by our consumption habits. In more equal societies, consumption is less about status. In more equal societies, there is more trust. You are more likely to trust your neighbours, colleagues, employer, and government in Sweden than you are in Saudi Arabia. Overall, there is more social cohesion. People are less likely to think their society is corrupt or that everyone is in it for themselves. There is a greater appreciation of collective responsibility, which builds a more efficient system for collective decision-making. Ultimately, in more equal societies, there is more trust in governments: the institutions we have built and tinkered with to help make long-term decisions. High levels of social cohesion and trust in political leadership are foundations for planetary stewardship. A fragmented, distrustful global political system will fail to stabilize Earth.

In Nordic countries, people recycle more, eat less meat, and produce less waste. Even before the pandemic, the number of airline passengers had begun to dip. In more equal countries, business people act differently, too. According to Pickett and Wilkinson, businesses in more equal countries are more likely to respond positively to international environmental agreements. We have spent quite some time talking to the CEOs of companies such as Ericsson, Scania, Spotify, and IKEA in Sweden. The anecdotal evidence that we gathered through these conversations bears this out. These business leaders are deeply concerned and take a personal interest in sustainability issues in their firms. They emphasise that quality of life is a big draw when they want to attract young talent to new jobs.

Greater equality will make it possible to achieve planetary stewardship. Yet inequality is rampant and getting worse. Globalization is creating a race to the bottom. Companies are stashing cash in offshore havens to avoid paying tax. Countries bend over backwards to attract wealthy businesses and turn a blind eye to tax avoidance. There is a general sense of political turmoil sweeping the world. How can we ever create more equal societies amid this chaos?

Never waste a good crisis, as Winston Churchill is alleged to have said. Roosevelt's New Deal emerged in a time of crisis. The Second World War led to a complete overhaul of international politics. In a few short years, Churchill and others created the International Monetary Fund (IMF), the World Bank, the forerunners to the

United Nations, and the World Trade Organization. These institutions have contributed to 75 years of relative peace, stability, and, therefore, economic development here on Earth. In 2020, the global COVID-19 pandemic was the biggest shock to hit our planet since the Second World War. As economies went into free fall, political impossibilities, such as trillion-dollar bailouts for companies and the distribution of millions of cheques to families across the United States, suddenly became possible.

Crucially, several ideas to tackle inequality and sustainability are emerging simultaneously. The Green Deal is making waves. The European Union is laying out plans to reach net zero emissions by 2050. The COVID-19 recovery fund of EUR 750 billion explicitly links economic recovery with green jobs. In the United States, the new Joe Biden administration has announced its own Green Deal to reach net zero emissions by 2050 and to hit 100 per cent renewable electricity by 2035. While there are many variants, in general these Green Deals are designed to bring together inequality and planetary stewardship. Through massive infrastructure investment in clean, green trains and mass transit systems in cities, power networks to distribute renewable energy, and energy efficiency systems, the Green Deals aim high. On top of that will come investment in health, education, and science. The Green Deals also plan to support workers in the coal industry, which will have to shut down.

Can any of this be achieved? Yes, it can. As the world reels and staggers from COVID-19 and economies look for ways to bounce back, it is hard to imagine a better time to implement Green Deals. The devastating impact of inequality is clear. Many leading institutions and media associated with unfettered free markets and rampant capitalism, such as the World Economic Forum, the *Financial Times*, and *The Economist*, support a radical overhaul of the system. This may surprise environmental and social justice activists. These institutions want to see a fairer distribution of wealth, a new social contract with society, and a future within planetary boundaries. The new leaders of the IMF and the European Central Bank feel the same way; they care about the future and fully understand what is at stake. The high priests of capitalism are demanding radical overhaul. It is possible that major transformation is imminent in the next decade.

Reducing inequality

There are three ways to reduce inequality. First, we could use taxes and benefits to shift money from rich to poor. Second, we could ensure a narrow difference in income before tax between rich and poor, by capping the salaries of CEOs to a fixed percentage above the average in the company, for example. Third, we could ensure that wealth does not accumulate much faster than economic growth. Since the Second World War, the Nordic countries, France, and Germany have taken the first route. Meanwhile, Japan has taken the second. The problem with the third route is that wealth moves internationally, often towards low tax regimes. To address this issue, Piketty has proposed a progressive global tax on capital: a wealth tax. Some argue that a wealth tax will slow essential innovation. This is simply not the case. All research suggests a wealth tax on highest earners has zero impact on innovation and profitability.

There are two more radical ideas that deserve more prominence. The first involves ditching our old thinking about taxes. The planetary emergency and gross inequality demand a seismic shift in approach. Instead of taxing good things, like hard work, we should tax bad things, like carbon emissions and destruction of nature. At the very least, we should tax high-polluting activities and remove tax for those on low incomes. In this world, those on low incomes would pay no tax, and flying would become very expensive. These changes must be done with care, as people tend to be loss averse: the pain of losing is greater than the pleasure of gaining. The French president found this out when he jacked up the price of diesel.

The second approach is perhaps not that radical at all, given the spending spree that suddenly popped up to prevent economic collapse in 2020. Governments, and the private sector too for that matter, should borrow more money and invest it in our future: social and physical investments in education, health, science, electrifying roads, battery storage, hydrogen economy, and so on. One way is by funding massive infrastructure projects and supporting Green Deals.

Why is borrowing for green investments such a good idea?

There are two main reasons. First, interest rates are at rock bottom and have been for a decade. There has never been a better time to borrow money. Second, there is a lot of money floating around. United States corporations (think Apple, Alphabet [parent company of Google], Facebook, among others) are sitting on USD 4 trillion in cash. The head of the major investment firm Berkshire Hathaway, Warren Buffett, tells everyone and anyone who will listen that he has USD 130 billion in cash burning a hole in his pocket. Where are the big transformative visions that require deep pockets but will build the pathways towards a safe operating space? The high-speed railways, bridges, offshore wind farms, and energy storage systems are the investments that will provide the pension funds with strong returns, 50, 75, and 100 years from now.

An even better approach than borrowing from the wealthy is, of course, to tax them, as Piketty points out. Perhaps we can implement both options at the same time. Win-win.

Let's recap. We live in a remarkable age. As Max Roser, the founder of Our World in Data, points out, if newspapers were published once every 50 years, the headlines today would read, "90 per cent of the population lifted from extreme poverty". Inequality is more than poverty eradication, though. The aim must be a fairer distribution of wealth in the world. This will build trust, moderate consumption, and help collective decision-making.

Some of the big picture solutions – such as a tax on wealth, a global perspective on tax, and a cleverly designed carbon tax – could, if managed well, be wildly popular among voters and help stabilize the planet. Green Deals are springing up around the world. We need to applaud and encourage them. We need them to become contagious. This will ensure long-term investment for a more equal, just world. But income inequality is unlikely to be solved in a decade, and we need to address other inequalities in society, too, not least relating to gender and race.

These are all generational challenges.

BUILDING TOMORROW'S CITIES

Note for Americans and other aliens:
Milton Keynes is a new city approximately
halfway between London and Birmingham.
It was built to be modern, efficient, healthy,
and, all in all, a pleasant place to live.
Many Britons find this amusing.

TERRY PRATCHETT AND **NEIL GAIMAN**
GOOD OMENS, 1990

Our friend Will Steffen tells a great yarn about his hometown, Canberra. Back in 2011, the Green Party won the local election on a promise to cut carbon emissions in half in a decade. They were duly elected, then admitted they didn't really have a clue how to do it (typical Greens, eh?).

But they succeeded anyway. Emissions more than halved. Since 1 January 2020, the city has run on 100 per cent clean electricity, the eighth city in the world to do so and the first outside Europe.

Buoyed by this success, and against a backdrop of fierce federal opposition, politicians now plan to make Canberra 100 per cent carbon neutral by 2045. The Greens plan to electrify the city's bus fleet and provide incentives to buy electric cars. They also hope to "Copenhagenize" transport by making walking, cycling, and scootering – micro mobility – the easiest, cheapest, and healthiest ways to move around.

So, how did Canberra achieve such alchemy? It was a no-brainer. The city is sunny. Renewables are cheap and efficient. Moving into renewable energy has brought jobs to the area. Farmers are more financially secure. Those suffering under prolonged droughts can now host wind turbines and solar arrays on their land. As Steffen says, "The biggest benefit of all will come in 2045, when we can look our children and grandchildren in the eye and say, 'We've done the right thing for you.'"

Here's to cities! The cause of, and solution to, all of life's problems.[14]

The fourth Earthshot system transformation is cities. They are the action arm of civilization. If you want something done fast, then look here. Cities are dynamic engines of innovation, creativity, and ideas. They create enormous wealth and are magnets for talent. They are the seats of power, where inhabitants feel closer to that power. Indeed, the mayor's office or town council is often just a walk or short bus ride away. This closeness is part of what gives cities their superpower.

Cities are also where we find crime and pollution, poverty and disease. They have colossal appetites. Cities cause 70 per cent of the world's carbon emissions. Not all cities emit equally, though. Of the thousands of cities in the world, 18 per cent of global emissions come from just one hundred of them.

Urbanization and industrialization are seen as the alchemy from which springs modernization and all-important growth. This is borne out by a simple statistic. Although you might expect a city's GDP to double as it doubles in size, this is not the case. In most places, the economic growth is often far more than double. Economic activity and a general fizz of creativity and innovation increase exponentially as more social connections are made and networks become increasingly complex.[15] Cities, with their combination of chaos and order, as well as social (people), biological (life), and physical (factories and cycle paths) factors, are complex machines that drive reinforcing feedback loops. Some of these feedback loops are good and encouraged, such as creativity and knowledge production. Some are bad, such as crime and disease.

---

14 This quote is originally by Homer Simpson. We have replaced the word "alcohol" with "cities".
15 This is another example of emergent behaviour.

We want to take you through three key insights on cities and how they need to transform. First, cities are on the front line of the Anthropocene. This decade will be their greatest test yet. If they do not adapt and evolve, they will shrivel and die. It is as simple as that.

Second, cities are superorganisms for responding to new information: flexing, rejuvenating, regenerating. Out with the old, in with the new. The coming onslaught of heatwaves and flooding provides extra incentives for cities to transform, and soon. To thrive and prosper, they need to attract the brightest, most creative people. The competition is tough. It turns out cities that aim to live within planetary boundaries are also great places to live. They are clean and pollution-free. People live without horrific congestion. The population is healthy.

The third insight is that cities can reinvent themselves in truly awe-inspiring ways. Never underestimate a city under pressure. Now, faced with climate change and other planetary threats, they are organizing themselves. Initiatives such as the Covenant of Mayors and C40 cities – two international networks – are uniting thousands of cities and local authorities.

Can you kill a city?

Cities are resilient. Anyone lucky enough to visit Jerusalem, Rome, Athens, or Istanbul may get lost in their richly layered histories stretching back millennia. Somehow, even when faced with cascading calamities, cities bounce back stronger than ever.

It is surprisingly difficult to kill a city.[16] People have tried really hard. Bombing campaigns in the Second World War reduced cities such as Dresden and Coventry to rubble. US and Japanese aircraft dropped payloads of bombs over Manila in the Philippines, smashing the city to smithereens. Most infamously of all, the United States dropped nuclear bombs on Hiroshima and Nagasaki. Who would have thought any of these cities would survive? But survive, thrive, and grow they did. Cities are superorganisms that can seemingly live forever.

---

16 By comparison, it is relatively easy to kill a company. The average age of a company in the S&P 500 index is 20 years. In the 1950s, it was 60 years, according to Credit Suisse.

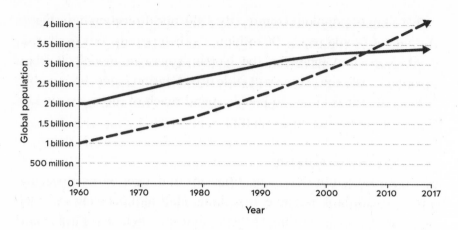

An urban planet
In 2007, the world crossed a
milestone: more people lived in
urban areas than in rural areas.

⟶ Rural population

⟶ Urban population

Over half the world's population[17] now live in urban areas. The
number of people living in cities is expected to increase by 2 billion
by 2050, bringing the urban population to 70 per cent. China is
driving the largest mass migration experiment in human history,
and some time in the next two decades its urban population is likely
to hit 1 billion people. Across China, India, and Africa, new cities
are being built to cope with this influx. The rate of urbanization is
truly astounding. The world needs the equivalent of a new
1.5-million-person city every week. Currently, there are 33
megacities, where the population exceeds 10 million. In 1970, there
were only two: Tokyo and New York. By 2030, a further 10 cities will
swell the ranks of the megacities. What this urban growth means is
that more than 60 per cent of the land projected to become urban
by 2030 is yet to be built. This is a blank canvas, a perfect opportunity
for transformative design.

We should feel a little uneasy about urban growth and its balance
between chaos and planning, though, as we step deeper into the
Anthropocene. In 1960, Lagos in Nigeria was a small coastal town.
Now, it is home to nearly 20 million people, the bustling hub of

---

17 55 per cent to be exact.

Africa's most populous country. By 2100, one projection indicates its population could reach 100 million. As their populations swell, Asia and Africa are following the North American model of urban sprawl. This is the least sustainable urbanization model imaginable, where the car is king and people spend much of their day stuck in traffic and choking on diesel fumes.

Our cities are sinking

There is no doubt that some cities are in danger of disappearing. New Orleans, for example, is sinking. Half of it is below sea level, and the ocean is eroding the surrounding wetlands. Levees and sophisticated flood barriers surround the city, making it into a modern-day fortress against the tide. The US Army Corps of Engineers designed the USD 14 billion project to provide protection against once in one hundred-year catastrophes in 2018. Less than a year later, they announced that by 2023 the defences will no longer provide this level of risk protection.

"Climate change is turning that 100-year flood, that 1 per cent flood, into a 5 per cent flood or a 20-year flood," Rick Luettich, a storm surge expert for the area, told *The New York Times*.

Based on a special report from the Intergovernmental Panel on Climate Change in 2019, it is likely that the ocean will rise between 0.39 metres (1¼ feet) and 1.1 metres (3½ feet) by 2100, depending on how much Earth heats this century. But the report also warns that city planners, engineers, and those with a low tolerance of risk might want to plan for a rise of a full 2 metres (6 feet) by around 2100 if Antarctica and Greenland collapse faster than computer models predict. They are already collapsing in line with worst-case scenarios. But even this does not capture the extent of the problem. The end of the century is only 80 years away, and sea levels will not stop rising in 2101. Coastal cities are locked into multi-century or millennial management. As we mentioned earlier, the last time temperatures soared to today's levels and held for a prolonged period, the sea level rose about 6 to 9 metres (20 to 30 feet). We need to be planning for this scenario, even if we manage to keep the temperature well below 2°C (3.6°F) global warming. There is a lot at stake. One billion people currently live on land that is less than 10 metres (33 feet) above sea level. This is why the next decade is so critical. We have wasted too much time already. But swift action now should allow us to slow the rate of rise, giving cities some breathing space to adapt.

New Orleans is a small city. The same fate awaits giants like Bangkok, Kolkata, Manila, New York, and Shanghai. Known as the "City of Joy", Kolkata is the cultural capital of India. The city of 14 million people lies low on the north coast of the Bay of Bengal. As early as 2050, it is at risk of annual flooding. Jakarta, in Indonesia, faces an even greater threat. It is sinking faster than any other city in the world. Built on a swamp connecting 13 rivers, parts of the city sank 2.5 metres (8 feet) in 10 years. Incredibly, this has not deterred property developers, who continue to erect luxury apartments along the coast. The problem in Jakarta is complex. Sea level rise and resource extraction both play a role. Groundwater beneath the city is pumped out to provide drinking water. This is a big driver, for now, of the subsidence.

Soon, a huge tidal barrier will surround the historically water-logged Venice. But as this nears completion, scientists are proposing even more ambitious ideas to protect New York and European cities. Recently, an idea to build a barrier around the entire North Sea was proposed: from Scotland to Norway, from England to France. You might imagine another between Spain and Morocco to protect the Mediterranean. Is this fortress world our future?

While floods rise from below, the heat from above is crippling cities. The smoke from the fires around Sydney in 2020 caused a hazy pollution that trapped heat and drove temperatures to a scorching 48.9°C (120°F). Six months previously, 19 million people in Delhi sweltered under 48°C (118°F) of extreme heat. A common narrative now is to say that this is the new normal. It is not. There will be no more "normal". We should expect constant turmoil spiralling upwards, whether it be the ocean, temperature, disease, or pollution. It is perhaps no surprise, then, that cities are mobilizing and innovating in order to evolve and thrive in the Anthropocene.

Superorganisms adapt or die

More than 600 cities have issued planetary emergency declarations, often focusing on climate. Copenhagen plans to be the first carbon-neutral capital city in 2025. Oslo aims to slash emissions 95 per cent by 2030. More than 7,000 cities from 133 countries have pledged powerful action on climate change. In 27 cities across the globe, emissions have fallen at least 10 per cent in five years. At the same time, their economies grew 3 per cent on average; this is viewed as robust growth in wealthy nations.

Let's look more closely at Copenhagen. Its audacious vision to become carbon neutral began in 2009. The city planned to reach net zero in just 16 years.

Here's what it takes for a small northern European city to reach that goal:

- 100 new wind turbines
- 20 per cent reduction in both heat and commercial electricity consumption
- 75 per cent of all journeys to be made by bike, on foot, or by public transport
- biogasification of all organic waste
- 60,000 square metres (646,000 square feet) of new solar panels
- 100 per cent of the city's heating requirements to be met by renewables.

The Copenhagen municipality now wants to encourage people to adopt a planetary health diet, too, to reduce the carbon footprint of the food they eat. They are working with our colleagues to understand how best to achieve this. A planetary health diet would both tackle obesity and encourage a healthy lifestyle. Copenhagen is the first city to do this.

So far, the city has slashed emissions 42 per cent since 2005. Over that time, its economy has grown 25 per cent. Between 2014 and 2015, emissions fell 11 per cent, beating the Carbon Law by some measure. According to Copenhagen's former mayor Bo Asmus Kjeldgaard in 2019, "We [the local government] combined life quality with sustainability and called it 'liveability'. We succeeded in building a good narrative around this, one that everybody could believe in."

The city's leaders are thinking in terms of systems. Take transport, for example. Copenhagen is offering citizens and visitors a single digital subscription service that makes it easy for users to travel by multiple modes: bus, metro, train, bike share, car share, and taxi. In

Nordhaven, a new development in the city, urban planners are creating the "five-minute city". They are minimizing car traffic by making shopping, day care, and other essentials all readily available within a five-minute walk. Of course, having the country on your side does not hurt. Denmark has plans to be independent of fossil fuels by 2050. If Copenhagen succeeds by 2025, we think the Danish transformation will come significantly earlier than this. Maybe by 2040 or even 2035.

A rich northern European city where the population demands high environmental standards is one thing, but surely poorer, smog-drenched cities in Asia cannot afford this type of transformation? Wrong. There are more than 425,000 electric buses on the roads in China today. Indeed, Chinese cities account for 99 per cent of the global fleet of electric buses. Opposite Hong Kong on the Chinese mainland, Shenzhen, a city of 12 million people, operates the largest electric bus fleet in the world. With more than 16,000 vehicles, the city's buses are 100 per cent electric. Taxis will soon follow suit. If you happen to visit a university in Beijing in the early morning, when students are hurrying to lectures, what really strikes you is the silence. Why? Well, transport is now dominated by electric mopeds.

All this makes a lot of sense. China's city dwellers endure horrific air pollution. Incredibly, 12 per cent of deaths in China are caused by indoor and outdoor air pollution. Globally, up to 9 million deaths every year are caused by air pollution.[18] More than 1 billion people in China are exposed to unsafe air for more than six months every year. Replacing traditional fossil fuel-powered transport with electric transport will reduce emissions and save lives. Without fossil fuel emissions, global average life expectancy would increase by one year.

Cities will change much faster than most predictions show. They are on exponential not incremental trajectories, and we have reached an inflection point. Cities like Copenhagen are showing the world that transformation is good for innovation, health, jobs, pollution, happiness, and well-being. Other cities will not want

---

18 Globally, death rates from air pollution have been falling, but this is primarily due to improvements in indoor air pollution. Death rates from outdoor air pollution have risen.

to be left behind. In 2019, when we were in New York for the United Nations Climate Summit, we found the most efficient way to get around the city was by bike. Hiring a bike is cheap, simple, and fast. We sailed through the gridlock, in contrast to previous visits when we dreaded even thinking about travelling around the metropolis. We imagine that a dense, compact city like New York could easily radically transform itself to a post-car innovation engine overnight. Imagine how much time everyone would save by ending congestion.

Not every city has the resources of New York or Copenhagen. Timon McPhearson, a professor of urban ecology at The New School in New York, reminds us that there are approximately 1 billion poor in the world, many living in the 1 million slums found across 100,000 cities. When poor people migrate from the countryside, they often cannot afford city rents. Instead, they build ramshackle temporary shelters without permission and with no rights to the land. Families have few incentives to invest in their homes if they could be bulldozed at any moment. Yet many dwellings survive for decades – occupied by the same families. In addition, many people living in these slums have an income and contribute to a bubbling economy.

The future of our cities

From all of this, we can identify four priorities for the Earthshot's system transformation in cities.

First, we need to ensure that everyone has access to clean drinking water and proper sewage systems. More people in the world now have mobile phones than access to toilets. Given the importance of slum dwellers to city economies, the most obvious solution is to offer property rights to them. This would provide security of tenure while simultaneously starting to build the infrastructure, including water, sewage, schools, and apartments, to provide hope, health, and dignity to the people who make a city thrive.

Second, planners need to eschew urban sprawl in favour of compact, efficient, green cities. Compact cities are ideal for district heating systems that can have zero emissions. Trees soak up pollution and keep pavements cool on punishingly hot days. We need to build sponge cities with large swathes of green parklands to soak up water instead of using concrete channels to drain it away. The cities could harvest this water for drinking and urban farming.

The third priority is efficient mobility. Most emissions from transport come from short trips around cities made longer by congestion. Let's kill congestion. Congestion in cities makes life miserable. Even people living in wealthy areas such as Silicon Valley and Los Angeles sit in traffic for hours a day. Economically, this is insanely inefficient and reduces quality of life. We do not need more cars on the road, even if they are electric.

Finally, the cities of the future will need to embrace circularity and regeneration and act more like living superorganisms.

The starting point is for cities to set targets for all planetary boundaries. Right now, we are working with the Global Commons Alliance, the World Economic Forum, and other organizations on a way to create such city targets based on the best science. For example, in April 2020, Amsterdam's deputy mayor, Marieke van Doorninck, announced that the city is planning to adopt the "doughnut" economic model based on planetary boundaries (see Chapter 16) – the first city to do so. The philosophy behind it is both profound and a no-brainer: the economic activity in the city should meet the basic needs of everyone (the inner ring of the doughnut) and it should do this within the means of the planet (the outer ring). Van Doorninck believes adopting doughnut economics will help the city to recover from the pandemic. We would argue it will make Amsterdam more resilient to the next crisis.

————————————————

Here's a curious emergent property of cities: people who live in cities have fewer children. Living in a crowded apartment with a family of 10 is uncomfortable. And in cities, women have more economic choices. These are two of the reasons why urbanization helps reduce population growth. And it does so without draconian, unwelcome one-child policies. In Chapter 14, we will address our fifth system transformation: population and health.

THE POPULATION BOMB DEFUSED

The number of children is not growing any longer in the world. We are still debating peak oil, but we have definitely reached peak child.

HANS ROSLING, TED TALK, 2012

Every day, someone somewhere will tweet us to remind us that we are wrong about what is causing the planetary emergency. Not just a little bit wrong. But very wrong in all possible ways. The gist of their attacks is this: "It's population, stupid."

Their argument is very simple. The exponential population explosion is the real emergency. If we do not tackle it now, then population will hit 100 billion very soon. But here's the thing. The countries stomping over Earth with the biggest footprints have the lowest birth rates. If everyone on Earth consumed resources like the average person in the United States, we would need an additional three to four Earth-sized planets with similar biospheres. Places with high birth rates, such as Niger, Uganda, and Mali in Africa, have some of the lowest environmental footprints on Earth and the worst poverty. Yes, high birth rates in these countries are unsustainable in the long run, as is grinding poverty. But it is wrong to blame the poorest of the poor for smashing through Earth's boundaries.

The fifth system transformation is population and health. We live on a crowded planet. As we know, a single cough in an Asian market can turn into a global pandemic within a matter of weeks.

Incredible progress in human health has led to an explosion in human population in the past century. Now, we face new health risks such as chronic obesity and pollution. These can threaten efforts to stabilize Earth and reverse the impressive progress already made. Climate change is also driving new health challenges: living and working in 50°C (122°F) cities, undernutrition as crops wither and die, and the spread of diseases like malaria to new areas.

The ultimate goal is to flatten the population curve. All the data show that we are approaching "peak child", as explained in Chapter 4. There are no data to suggest that the population will keep rising at the high rates seen 30 or 40 years ago. If Hans Rosling were still with us, he would undoubtedly bemoan that people have not updated the knowledge they learned decades ago at school. Back then, population growth did, indeed, appear to be out of control. If the population were still increasing at the same rate it was between 1980 and 2000, then it would reach a staggering 540 billion by 2300. But infinite exponential growth is not possible in the real world; instead, everything eventually slips into an "s curve", as growth rates slack off. This is the pathway we are on.

At the start of the Holocene, a mere 5 million people are estimated to have roamed Earth. Eventually, this increased to approximately 1 billion by about 1800. This is a glacial growth rate of less than 0.1 per cent a year. During the 20th century, the population rose from 1.65 billion to 6 billion. Since 1970, the number of people on Earth has more than doubled. The global population is now topping 7.8 billion and is expected to hit the next milestone of 8 billion in the early 2020s.

For most of our history, women gave birth to between five and seven children. They expected half of them to die before they were old enough to have children themselves. As Rosling points out: "We did not live in harmony with nature; we died in harmony with it." We lived under a constant threat of famine, disease, and war. Heartbreak and misery were common for most people until the industrial revolution. It then took a few generations for our minds to adjust to the new reality of genuine food security, healthy diets, effective medicine, and, since 1950, low levels of war or conflict.[19]

The population growth rate reached a maximum of around 2.1 per cent per year in the 1960s. Today, it is at 1 per cent, and

Population growth

The population growth rate has plummeted since the 1960s. This is one of the biggest achievements of the past 50 years.

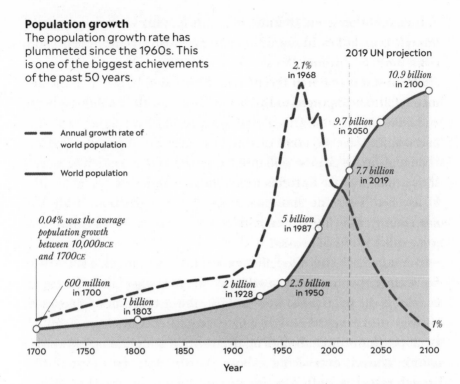

2019 UN projection

2.1%
in 1968

10.9 billion
in 2100

- - - Annual growth rate of
world population

——— World population

9.7 billion
in 2050

7.7 billion
in 2019

*0.04% was the average
population growth
between 10,000BCE
and 1700CE*

5 billion
in 1987

600 million
in 1700

2 billion
in 1928

2.5 billion
in 1950

1 billion
in 1803

1%

1700   1750   1800   1850   1900   1950   2000   2050   2100

Year

continues to fall. This global figure hides a startling divergence. In Japan and South Korea, the number of children per woman is about 1.4 and 1, respectively. In Niger, Africa, that number is 6.9. Remarkably, Hong Kong's fertility rate was lower than China's during its one-child policy (1979–2015). The high rates of female education, high population density, and high cost of having children meant that many people simply felt they could not afford more than one child.

Yet, in the 21st century, population continues to grow strongly. Not nearly as strongly as before, but it is growing. This is because people are living much longer. A child born in 2016 could be expected to live 72 years. A girl might expect to live for 74 years and a boy about 70. This is the global average, not the average in the wealthy countries of Europe or North America. Forget the Moon

---

19 Steven Pinker's *Enlightenment Now* (2018) and others show conflict has indeed reached the lowest point in history.

landings – this combination of science, policy, and economic strategy is arguably humanity's most impressive achievement, and it continues to this day. We should be rejoicing in the streets.

The progress made in recent decades is mind blowing. Someone born in 2016 is expected to live 5.5 years longer than someone born in 2000. This increase in global average life expectancy was the fastest since the 1960s and reversed the declines seen in the 1990s, when the Soviet Union collapsed and the AIDS epidemic swept Africa – economic fortunes are tightly linked to human health. Africa has seen the most astounding transformation, with life expectancy rising by 10 years. Someone born in Africa today can now expect to live 61 years.

In many rich nations, life expectancy continues to extend. However, this is not the case everywhere. In some places, the trend is reversing. In the United States, life expectancy plateaued in 2014 at 79 years and is now decreasing, due to a heap of social problems, including drug overdoses during the opioid crisis, suicide, obesity, and alcohol abuse. This news comes despite the United States spending more per capita on health than other major economies.

Globally, longevity is expected to increase as poorer nations develop rapidly. This means that the best population estimate for 2050 is about 10 billion people. By 2100, it may have climbed to 11 or 12 billion, but probably no more. This could change dramatically if poor countries develop faster. As industrialization kicks in, fertility rates will fall significantly. Investment in education for girls and women cuts rates dramatically, as does migration to cities. Focusing on these two factors could help ensure that population stabilizes at a lower level.

A crucial question is: how many people can live on Earth within a stable biosphere? In the past few decades, researchers have puzzled over this and failed to agree. Some say that Earth cannot support even 1 billion people sustainably with a high quality of life. Others have pinned an upper limit at a staggering 98 billion people. This wide range is a little unhelpful. Our own research shows that it is possible to feed a population of 10 billion within planetary boundaries. However, we will need to transform the farming system, ditch junk food in favour of a healthy diet, and cut food waste.

Interestingly, there are no studies showing that our current unsustainable world can host 10 billion people while meeting minimum living standards. We will simply run out of land, natural resources, air, water, and biodiversity, while triggering catastrophic climate extremes. The Earthshot is the only pathway that has a chance of delivering these standards for a growing world population.

Hosting 10 billion people on Earth will be extremely challenging, though. So far, no country operates within planetary boundaries. Adding another 2 to 3 billion people will require us to succeed with all six system transformations. For food, in order to feed 10 billion people, we will need a global transformation to sustainable agriculture; a transition to more plant-based diets; additional investments in alternative sources of animal protein, such as lab-grown meat and synthetic food products; biotech advances; and sustainable aquaculture. Go beyond 10 billion, and all this becomes even more daunting.

But even as we make space for another 2 billion people on our planet, we face many other health risks in the Anthropocene. Air pollution now kills up to 9 million people annually. Antibiotic resistance is another looming tragedy of the global commons. Antibiotics help individual farmers manage disease in their livestock, but their overuse drives rapid evolution of resistance. Now, some bacteria are resistant to the most powerful antibiotics and we risk exceeding planetary boundaries for antibiotic resistance. And pesticides damage both human and environmental health. Arguably, though, the biggest risk is a global pandemic.

Fortunately, science is better prepared than ever to untangle the genes of deadly diseases. Just 10 days after the first reported case of COVID-19, scientists published the gene sequence online to help other researchers to begin the hunt for a cure. However, the economic foundation of the health system in the world is broken. Pharmaceutical companies are more interested in making new opioid painkillers or cures for baldness than investing in antibiotics or cures for potential diseases that have not yet emerged. As Bill Gates notes, "Government funding is needed because pandemic products are extraordinarily high-risk investments."

COVID-19 has shown that as a global civilization we are dangerously exposed to abrupt shocks. Cooperation among countries and colossal efforts to shut down economies effectively contained the disease in many places, while tragic political incompetence fuelled distrust elsewhere, allowing the disease to rip through towns and cities. The pandemic has taught us that health is a public good. The good health of all citizens benefits everyone. Yet, we are undermining this public good.

Tackling health and population is essential for the long-term viability of a global civilization within a safe operating space. To summarize, we need a radical overhaul of our global health system, from surveillance and monitoring for disease outbreak to scaling-up of vaccine production, reducing air pollution, tackling obesity, and managing growing resistance to antibiotics. This sounds like a lot – it is a global system transformation, after all – but success brings win-win situations. Reducing obesity and improving diets can bring down greenhouse gas emissions. Reduced greenhouse gases will lower the levels of air pollution. And providing family planning and education to girls has the potential to avoid 85 billion tonnes (93 billion tons) of carbon dioxide emissions this century and to stabilize global population at levels that are manageable.

Ultimately, we have crossed the demographic tipping point. The problem is not the extent of future population growth in Africa and South Asia, though. Rather, it is the lifestyle choices made by the growing middle classes and wealthy in poorer countries. That said, it is a moot point whether "choice" is even a realistic possibility in a world of inequality, coupled with laser-guided marketing and advertising designed to target our insecurities in order to fuel consumption. If those of us emerging from poverty aspire to the lifestyles of those of us driving SUVs and taking long-haul flights to post selfies on social media from exotic locations, then the Earthshot mission will miss its target by a long way. If, on the other hand, the wealthy nations of Europe, the United States, Japan, and elsewhere transform to healthy, happy societies fuelled by vibrant, regenerative, circular economies with zero carbon and zero nature loss, then this sustainable societal model will become the aspirational North Star.

TAMING THE TECHNOSPHERE

We invented fire, repeatedly messed up,
and then invented the fire extinguisher,
fire exit, fire alarm and fire department.
We invented the automobile, repeatedly
crashed, and then invented seat belts, air
bags and self-driving cars. Up until now,
our technologies have typically caused
sufficiently few and limited accidents.

MAX TEGMARK
*LIFE 3.0: BEING HUMAN IN THE AGE OF
ARTIFICIAL INTELLIGENCE, 2017*

"One small step for man, one giant leap for mankind." On 20 July 1969, Neil Armstrong and Buzz Aldrin set foot on the Moon. My parents listened to the Moon landing on the radio in a car park overlooking the river Shannon on the west coast of Ireland. My mother was pregnant with me at the time.

The Moonshot affected everyone. Like many kids, I wanted to be an astronaut when I grew up. I loved maths and physics and studied spacecraft engineering at university. Looking back, I realize that engineering training is very different from scientific training. Scientists look for problems. Engineers look for solutions.

Like many, once I was a techno-optimist: all we need is more investment in more innovation and this will drive the change we need. I imbibed the intoxicating narratives coming from tech CEOs and other evangelists. Soon the world's information will be organized and people will be connected to one another, promoting a shared world view based on science. This has not panned out quite as planned. Yes, we need innovation, but innovation is what has led to a destabilized Earth. Innovation must now be directed. And on its own, it is not enough. The planetary emergency demands behavioural change and political and economic transformation, too. It is a full systems change, and technology can drive it. **Owen**

———————

Technology, particularly the digital revolution, is the sixth and final system transformation. Technological innovation is not neutral. In our current economic system, control lies in the hands of those with the deepest pockets, many of whom have scant regard for a stable planet. The richer we become, the greater the impact of our technology on the stability of Earth. We need to break this vicious cycle. Technology must now be harnessed to stabilize Earth.

One thing we can be sure about. Unlike the other system transformations, which will need almighty efforts to change course, technological disruption in the next few decades is assured: machine learning, artificial intelligence (AI), automation, and the Internet of things are hurtling towards us. The direction these technological innovations take in the next decade will determine whether we succeed in our Earthshot, or whether we crash and burn.

Humanity has created a lot of stuff, sometimes called the technosphere. We have made enough plastic to wrap Earth in cling film. We have mixed enough concrete to make an exact replica of Earth 2 millimetres (0.08 inches) thick. The physical stuff we have built weighs more than 30 trillion tonnes (33 trillion tons). The waves carrying our first radio and television broadcasts have travelled more than 100 light years from our solar system, past unfathomably distant stars where planets a bit like our own orbit – although the signals will be too weak for any intelligent life to detect. Perhaps it is time to direct more of this technological know-how into the Earthshot.

Money is already flowing towards a clean, green future: energy generation; electric cars, trucks and even aircraft; more efficient buildings, and alternatives to steel, concrete, and aluminium manufacturing. The speed of change is phenomenal. For example, South Australia was hit by a series of blackouts in 2016–17 and needed a solution. Eccentric billionaire Elon Musk made a very public bet on Twitter that Tesla could build a giant battery in Australia in 100 days or he would pay for it himself. And lo, the world's largest grid-scale battery storage system emerged in the desert. Costing about USD 100 million, it has stabilized power in the region and saves about USD 40 million a year. Visible moments like these create investor confidence. Sentiment is swinging from scepticism five years ago to wholehearted embrace now. And as we will see, economic sentiment and confidence are crucial factors when it comes to technological acceptance and diffusion in society.

It may come as a surprise to hear that we already have the knowledge and most of the technology to live within all nine planetary boundaries. If solar and wind keep scaling as they are today, they will supply half of our global electricity by about 2030. The trick is to ensure that the technology keeps diffusing, by remaining on strong price and innovation curves. Sure, some new technologies will be essential: creating a hydrogen economy and zero emissions aircraft, for example, will require huge investment in innovation. We also need technologies to adapt to the change that is inevitable to protect societies from rising sea levels, extreme events, and the spread of disease. Beyond that, given the planetary crisis we have now entered, we must also consider geoengineering: our last-ditch attempt to restabilize Earth. Most of all, we somehow need to rein in the technologies that work against these goals.

What makes technology diffuse in society?

Some new technologies spread exponentially, while other great ideas fail to gain momentum. What is the secret to a technology's success? Exponential technologies tend to be granular rather than big and bulky. Think smartphones, not nuclear power plants. The technology needs to have the capacity to evolve rapidly and diffuse through society, which means that it needs to be relatively cheap, too. Exponential technologies are a combination of technology, economic dynamics, and social dynamics. A nuclear plant is, to state the obvious, big, static, and expensive. Innovation is slow because each generation of plants lasts decades. Solar and wind

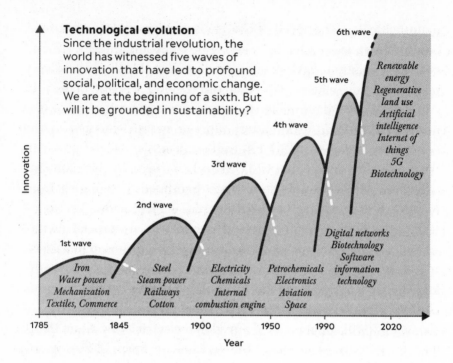

Technological evolution
Since the industrial revolution, the world has witnessed five waves of innovation that have led to profound social, political, and economic change. We are at the beginning of a sixth. But will it be grounded in sustainability?

Innovation

1st wave

*Iron
Water power
Mechanization
Textiles, Commerce*

2nd wave

*Steel
Steam power
Railways
Cotton*

3rd wave

*Electricity
Chemicals
Internal
combustion engine*

4th wave

*Petrochemicals
Electronics
Aviation
Space*

5th wave

*Digital networks
Biotechnology
Software
information
technology*

6th wave

*Renewable
energy
Regenerative
land use
Artificial
intelligence
Internet of
things
5G
Biotechnology*

1785   1845   1900   1950   1990   2020

Year

power are the opposite. They are small, comparatively inexpensive, and scale quickly. The innovation is rapid. Solar and wind increasingly outcompete nuclear and fossil fuels. Both these older technologies have seen remarkably little innovation in decades. Nuclear, for example, was once seen as an essential technology for a clean future, but now its own future is in doubt.

Offshore wind power is probably the most significant breakthrough in recent years. Fossil fuel technology allowed us to get very good at building big objects like drilling platforms in remote locations such as the North Sea. Now, we can use this know-how to build offshore wind turbines. Even if we only build wind turbines in offshore waters that are shallower than 60 metres (197 feet), we could still supply more than enough[20] electricity to meet the world's demands.

But do we even need all the energy that we currently consume? Digitalization will radically affect how much energy we use. By

---

20 We would actually have enough electricity to power the world's current electricity demands plus an additional 50 per cent.

shaping the direction of technological evolution towards doing more stuff with fewer devices – cramming more applications onto a single smartphone – and sharing more resources, from cars to office space and manufacturing, then demand for energy could fall by up to 40 per cent. This is a more disruptive idea than it first appears. Most mainstream energy analysts predict energy demand will grow, not fall, in the coming decades as nations develop.

Price is a critical part of the equation in technology diffusion. It is often overlooked, but Moore's Law (described in Chapter 10) is driven by both doubling of computer power and halving of price. This double helix of power and price drives phenomenal growth and innovation in a short period and opens up whole new markets. This can be seen for both hardware, such as smartphones or tablets, and software such as social media platforms. When innovators on these curves enter markets, they effortlessly outcompete incumbents. Their rapid innovation cycles are sensitive to what users want. This allows them to improve their new products rapidly as price falls. Music streaming destroyed record shops, Amazon came close to killing bookshops, and Netflix annihilated video stores. The speed and scale of disruption always comes as a surprise to industries.

This trajectory is already happening with wind, solar, battery storage, and electric vehicles. We have already reached a tipping point where the change is now unstoppable (see Chapter 18). We will become planetary stewards because it is the best solution at the cheapest price.

So, technology is on the side of planetary stewards? Not so fast. The grand challenge is still on our Earth watch. How do we make the digital revolution work within planetary boundaries? Moreover, how do we steer the revolution towards supporting and accelerating the Earthshot goal? And how do we do it in a way that avoids rebound effects: where digitally enhanced smart homes reduce emissions, saving money that is then spent on high emissions consumption elsewhere? The digital age has to work for the planet.

Technology's dark side

Technology companies are on the side of money. As oil gets more difficult to find and extract, oil companies have jumped into bed with Google, Microsoft, and Amazon to use AI and algorithms to search for the last remaining reservoirs. In 2018, oil companies threw about USD 1.8 billion at AI. By 2025, this may be as much as USD 4 billion. Needless to say, this is at odds with planetary stewardship.

But it gets worse, far worse. Technology has changed the way information flows in the world. We freely donate our data to digital platforms, such as Amazon, Facebook, and Google, revealing to them our innermost desires. These companies employ psychologists and neuroscientists who use our data to predict our behaviour and influence it. This information is on sale to the highest bidder. Without proper checks and balances, these companies are undermining our ability to distinguish fact from fiction. They are making society more polarized, which, in turn, is destroying democracy and ruining communities. Change is happening so fast that society cannot keep up.

The reason technology companies now command trillion-dollar market capitalization is because they have direct access to the inner thoughts and desires of the world's 4 billion middle-class consumers. Facebook, Google, and their ilk provide platforms for free: if you are not paying for the product, then you are the product. This is an extractive industry. But instead of extracting oil or coal, companies are extracting something far more valuable: behavioural data.

Marketing and advertising have always preyed on our desires, fears, and insecurities to drive consumption. Now, companies do this with laser-guided precision. Ironically, as many people call for a new approach to capitalism in order to stabilize Earth, a new form of capitalism is emerging, dubbed "surveillance capitalism" by US academic Shoshana Zuboff. Our freedom of thought is under attack and exposed to abuse by corporations and states. It is unclear whether this will help restabilize or further destabilize our planet.

I, for one, welcome our new computer overlords

Currently, we are on the brink of the next wave of technologies: AI, automation, machine learning, and the Internet of things.

In 2011, a computer took on the two best players of the US television quiz show *Jeopardy!*. Before the game, Ken Jennings had won 74 times in a row, and Brad Rutter had hauled in the biggest prize pot in history. IBM's Watson computer crushed them both. "I, for one, welcome our new computer overlords," said Jennings rather sanguinely on defeat. Watson's success relied on "deep learning" and natural language processing. It can find meaning in vast amounts of text. Our technology is becoming more powerful, and this power is increasing exponentially.

In 2017, DeepMind's AlphaZero achieved superhuman abilities in chess, Shogi,[21] and Go[22] within 24 hours. Critically, AlphaZero's handlers did not train the AI by dumping on it millions of games won by grand masters. Instead, they simply taught it the rules of the games. It figured out the rest itself.

Where will this lead us? Scientists are already using some deep learning algorithms to find new antibiotics. And researchers have sent an AI to crawl through millions of academic papers in search of possible new materials. But some caution is required as we step into this brave new world. Many of the people shouting loudest to tread carefully are the world's leading experts in AI. As Max Tegmark writes in *Life 3.0*, "Intelligence enables control: humans control tigers not because we're stronger, but because we're smarter. This means that if we cede our position as smartest on our planet, it's possible that we might also cede control."

Have we already ceded control to algorithms? Perhaps we gave up control to algorithms several generations ago. And we are not talking about Netflix recommendations. The global markets influence society, business, and politics. We are at the mercy of high frequency trading. The markets ebb, flow, dive, and soar according to simple algorithms – buy, sell – like a proto-consciousness driven by sentiment, mood, and herd instincts.

In the 2020s, automation will increasingly replace jobs across many industries. This is a major concern. Robotics and automation are predicted to displace 20 million factory workers worldwide by 2030. Robots will drive our trucks, trains, and planes, flip our burgers, make our lattes, and stock our shelves, if they aren't already. But it is not only low-skilled labour that is at risk. Management consultancy McKinsey estimates that much of a law clerk's job and even a lawyer's job can be automated. Automation is creating the

---

21 A Japanese form of chess.
22 A devilishly complex Chinese board game.

right conditions for a perfect storm of rising unemployment, deepened inequality as wealth is more concentrated to a few, rising insecurity, and rising resentment, which opens the door for more populist, authoritarian leaders. This has the potential to destabilize societies. We need to find a way to redirect our innovative capacity and at the same time build safety nets, such as retraining programmes, into society to protect people. Essentially, we need mission-driven not scattergun innovation. We'll address this more fully in Chapter 18, where we discuss policies.

Another technology predicted to break through in the coming years is blockchain: a decentralized way of storing information on transactions such as purchases or legal contracts. It has been touted as a major disruptor of everything and a perfect tool for carbon trading. Recently, Mercedes-Benz announced a pilot scheme to use blockchain to track carbon emissions in its cobalt supply chain. If blockchain does somehow break into the mainstream, it will be a disaster for stabilizing Earth in its current form. The blockchain application Bitcoin currently generates more carbon emissions than the Republic of Ireland, but unlike Ireland, barely anyone uses it.

The uncertainties around the direction of technology are profound. Forget about AI, automation, or blockchain. It is not even clear how 5G technology – the next generation of mobile networks – will disrupt industries, and it is being rolled out as we speak. For now, we are only told to look forward to playing online games with less delay and fewer glitches. So far, so benign.

In 2020, the BBC saw a 5G world beyond better gaming, "Imagine swarms of drones co-operating to carry out search and rescue missions, fire assessments and traffic monitoring, all communicating wirelessly with each other and ground base stations over 5G networks." We can equally imagine a world of drones used as part of a mass surveillance system. 5G will change how we interact with technology and how our devices and appliances speak to one another. This could pave the way for autonomous vehicles and more efficient cities. While one scenario is smooth-flowing traffic, another is a traffic system in which drivers automatically bid against each other at every intersection. The highest bidder wins the right to move first. What 5G will mean for privacy, security, and democracy is anyone's guess.

At present, governments are failing to steer this many-headed beast that is technological innovation in directions that will support societal goals. The tech companies themselves, such as Amazon, Google, and Facebook, talk a good game when it comes to reducing their own emissions, but this is a drop in the ocean compared with the impact of their users. Given their influence over the world's consumers, these companies must step up to this new responsibility.

Geoengineering

If all else fails, can we restabilize Earth using extreme technological fixes? In the worst-case scenarios, protecting billions of people will take unprecedented feats of engineering. Geoengineering aims to address climate change using deliberate and large-scale technological interventions. Think terraforming our own planet. To be honest, most of these ideas come straight from the pages of science fiction. But many are now getting serious scientific attention. By 2030, we should know which ones are our best bets.

Geoengineering comes in two flavours. The first option is to block sunlight reaching Earth. The second is to suck greenhouse gases out of the atmosphere. Both are insanely high-risk interventions in a complex system.

There are several ways to block sunlight, starting at a cosmic level. Erecting giant sunshades between Earth and the Sun would do the job nicely, by potentially stopping about 2 per cent of incoming heat from the Sun. The numbers have been crunched. We would need hundreds of thousands of 1-metre-square (10-foot-square) sunshades weighing approximately 18 million tonnes (19.8 million tons). In total, it would cost a few trillion dollars and last about 50 years. But this does not help ocean acidification, because carbon dioxide will still be building up in the atmosphere. If we keep emitting, even if we block incoming solar radiation, then the ocean will get steadily more acidic – one of the major causes of past mass extinctions. In addition to the cost and engineering challenge, giant sunshades are likely to bring unexpected consequences: for example, changing weather patterns around the globe.

Probably the most talked-about geoengineering solution is to dump millions of tonnes of tiny particles high into the atmosphere in order to reflect heat back to space. We know this works. Every major volcanic eruption ejects ash into the upper atmosphere. This has a measurable effect on climate. When Mount Pinatubo erupted in the Philippines in 1991, the planet cooled a little for the few years after the initial eruption,[23] but this impact was short-lived because these particles dissipate in the upper atmosphere within a few years. The scale of this type of intervention would need to be immense: some 3 to 5 million tonnes (3.3 to 5.5 million tons) of sulphur ejected every year.

Cloud seeding or whitening is another option. Churning up large tracts of the ocean can throw salt particles into the atmosphere that help cloud formation. More clouds will reflect more heat back to space and perhaps cool the planet. This idea could be used quite locally to protect coral reefs, for example. At a global scale, though, this would take vast armadas of autonomous ships plying the ocean for evermore.

We could also simply paint our roads, roofs, and cities white to reflect heat. Locally, this effect could keep towns and villages cooler. A similar proposal is to grow genetically modified crops that are better at reflecting heat from the Sun, for more widespread cooling. There is a dangerous thread running through all of these geoengineering ideas, though. Once we start we cannot stop. If we are forced to stop a geoengineering project for whatever reason – the money runs out, geopolitical strife, catastrophic unforeseen consequences, for example – then Earth's temperature would rocket abruptly.

Several ideas for sucking carbon dioxide out of the atmosphere have also been proposed. The most commonly talked about is carbon capture and storage. There are two main ways to do this. The first is to pull carbon out of the atmosphere using some kind of machine. The second is to grow and burn plants for energy. Burning the plants releases carbon dioxide, but this would then need to be trapped and put somewhere safe, away from the atmosphere. The most common proposal is to pump it back into used oil reservoirs, deep beneath the sea, for safekeeping. However, if we rely on plants to capture the carbon, the scale needed

---

23 0.5°C (0.9°F) for two years.

will interfere with global food production, and we will struggle to provide enough food for our growing population.

Ultimately, some of these technological solutions will be required, even if the world makes massive emissions cuts, because we are so close to unmanageable risks. When geoengineering becomes essential, we should plan for a smorgåsbord, and deep systematic assessments of risks. Carbon capture and storage seems the most promising option: it is economically viable and appears relatively safe. In the next decade, we need to start scaling it up, so that we are ready to pull 5 to 10 billion tonnes (5.5 to 11 billion tons) of carbon dioxide out of the atmosphere every year. We will need this even if the world follows the Carbon Law. Going further than this, though, really is in the realm of science fiction.

Finally, researchers have also proposed a way to stabilize parts of the Antarctic ice sheet. It would take about 12,000 wind turbines to generate the power, but giant snow machines could be employed to suck up sea water and turn it into snowfall to rebuild the ice sheet and protect the world from several metres of sea level rise. Our assessment is that ideas like this are, for now, interesting projects on paper and in the minds of brilliant colleagues. While they highlight the sheer scale of the challenges we face, they are perhaps unrealistic at the moment. Ten years from now, we might be revising this opinion. These are the extremes we are being forced to consider.

Needless to say, this is a race against time. Accelerating technological progress can help drive the other five major system transformations we desperately need. But it could equally hinder our chances of success if technology merely pushes emissions higher by increasing unsustainable consumption.

In the next two chapters, we will look at how economic and political policies can help guide the six system transformations that are the foundation of our Earthshot mission. In Chapter 18, we will return to tipping points once again.

A GLOBAL ECONOMY WITHIN PLANETARY BOUNDARIES

Today we have economies that need to grow, whether or not they make us thrive. What we need are economies that make us thrive, whether or not they grow.

KATE RAWORTH
DOUGHNUT ECONOMICS: SEVEN WAYS TO THINK LIKE A 21ST-CENTURY ECONOMIST, 2017

I recently caught up with an old friend, Jochen Zeitz, the former CEO of the sports giant Puma. Under him, Puma realigned itself with the planet. Now, he is CEO and chairman of Harley Davidson. Jochen told me that Harley has decided to go electric. I almost fell off my chair. Harley Davidson, the epitome of loud motorcycles, oil, and combustion engines is launching electric motorcycles. Is that even possible? Actually, it is about their survival. They have realized that the next generation will demand fossil fuel-free products.

If Harley can do it, so can the world. **Johan**

One of the first people to get to grips with what planetary boundaries meant for economic theory and practice was Kate Raworth, now an academic at the University of Oxford. Raworth realized that if the planetary boundaries represent the environmental ceiling for the global economy, then there is an equal and opposite social floor – sufficient access to energy, water, food, good health, education, housing, and more (12 in total). She called this new economic model the "doughnut" (see overleaf).

Doughnut economics is now beginning to enter economic orthodoxy. In 2020, the University of Oxford published a new economics textbook charting the evolution of economic thought, starting with Adam Smith in 1776 and ending with Raworth's doughnut model. Smith famously introduced two metaphors – capitalism's "rising tide raises all boats" and the market's "invisible hand" – to explain how self-interest, markets, and growth bring unintended social benefits for all in society. In the Anthropocene, the rising tide, nudged along by capitalism, is no longer a metaphor; it literally threatens to sink cities, with the poorest hit hardest, and the invisible hand is pushing people's heads underwater. Now the challenge is to redirect our economic power within the safe space for humanity. And as fast as possible. If the planetary boundaries are the fencing to prevent the global economy from tumbling off the cliff, we have already climbed over the fence and are leaning over the edge. We are not yet hanging on by our fingertips, but soon we will be. Then, governments will be forced to take over industries to attempt to pull us back to safety. But if we reach that stage, it may well be too late. This scenario is closer than most would like to admit. Perhaps the doughnut is more like a lifebuoy.

So, are there any reasons for optimism? We believe there are several beacons of hope indicating that we can live prosperous and equitable lives within planetary boundaries, all 10 billion or more of us. And we can build an economy that takes us there because anything that is technologically possible is economically possible. The economic system is our most powerful tool for transformation. The secret to success will be telling a more compelling economic narrative than our current status quo. As French author Antoine de Saint-Exupéry said, "If you want to build a ship, don't drum up the men to gather wood, divide the work, and give orders. Instead, teach them to yearn for the vast and endless sea." This is not such a difficult task. The global financial crisis of 2008/9 and the 2020 pandemic have already exposed the deep fractures and fault lines of the neoliberal narrative that has left its mark in the geology of our planet.

Reasons to be optimistic

The first reason for optimism is that the new economic logic is far more attractive than our current way of doing business: the return on investment on a stable planet, where economies adopt renewable and circular principles, is truly in balance with long-term rewards. The markets are beginning to catch on. Sustainable technologies and business models are proving to be more profitable, thanks to more efficient production and growing market demand. As CEO of Scania Henrik Henriksson notes, "Sustainability and profitability now go hand-in-hand." And he runs a heavy truck business.

Second, we have been here before. New economic thinking can become mainstream surprisingly quickly. In the 1930s and 1940s, Franklin D. Roosevelt's New Deal – the long-term investments and new social contracts between labour, government, and companies – pulled the world out of the Great Depression and helped rebuild economies after the Second World War. In the 1980s, the economic ideas of neoliberalism swept the world, driven by Margaret Thatcher and Ronald Reagan.

Third, we should remember, too, that we are not starting from scratch. The energy revolution began 30 years ago and it has now reached take-off speed. Population growth is slowing. The technology revolution is unstoppable, and cities are in a constant state of reinvention. We have a strong, healthy ecosystem for rapid iteration of ideas and diffusion. With 190+ nations and thousands of cities, we can test solutions and learn from one another, which is a good base for transformation and innovation.

And finally, the 2020 pandemic provides this generation's most important moment to reinvent the global economy.

At its heart, economic development in the Anthropocene means two things: reshaping the playing field that businesses operate within – the markets, and the flow of goods and services – and long-term planning for the future – cathedral thinking. Before we tackle the markets and cathedrals, first we need to deal with growth.

We are obsessed with growth

Politicians obsess about economic growth. It is used as a proxy for opportunity, stability, well-being, and happiness. If we have growth, we will have all of these things, too. But in other areas of life,

Doughnut economics
Raworth's economic model combines the
planetary boundaries framework with
social considerations such as
education, food, health,
and inequality as well
as access to energy,
housing, and water.
Within this model, many
economic pathways (inset)
are possible as long
as materials are
recirculated.

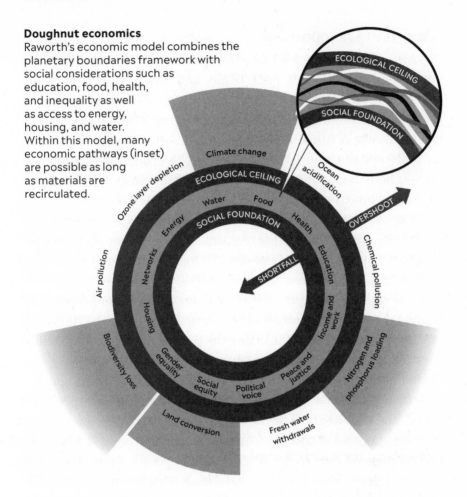

we do not associate never-ending growth with positive outcomes. In our gardens, we want plants to grow, but also to stabilize and stop at some point. We want our children to grow, too, but not indefinitely. As we saw in Chapter 13, greater wealth does not translate to greater well-being or happiness, even for rich people. But as we get richer, our environmental footprint expands. This is a general rule.

Where did this obsession with growth come from? The US economist Simon Kuznets first developed the idea of GDP in 1934. In the 1940s, despite Kuznets' own warnings that GDP failed to capture not only the true value of society but also the environmental and social damages inflicted on societies by the economy, it became the defining measure of a country's success. This is understandable: it is a simple and powerful indicator of a country's rise out of poverty.

For the past 200 years, the global economy has grown exponentially, defying all odds. But growth is often subsidized by cheap labour: exploiting poor adults and children in developing economies. And growth is subsidized by the planet: mining, soil erosion, deforestation, air pollution, and greenhouse gases. When you think about it like this, nothing is growing really. Social and natural capital are just being converted to economic capital. Net zero. There is no growth. The subsidized phase of consumer capitalism has come to an end. The free lunch is over.

Economic growth comes from a combination of investment in education, innovation, infrastructure, urbanization, and energy. Resource extraction is an important factor, as is international trade. And other things, too: political stability and trust in public institutions, for example, are critical for market confidence. Of course, we should not forget debt – borrowing from the future to pay for today. In early 2020, before the pandemic caused economies to crash, global debt reached a record USD 258 trillion.

Despite political obsession with growth, wealthy, technologically advanced nations have stopped growing at the breakneck speeds of the 1950s and 1960s. The slowdown occurred more than three decades ago, and according to many economists a return to high growth may not even be possible. There are good economic reasons for this. Developing economies have high population growth rates and expanding work forces as people move to cities in search of a better life, thus enabling rapid growth. Rich nations have low or zero population growth; they have largely urban populations who are highly educated and are more likely to work in sectors such as finance, technology, knowledge, health, hospitality, or legal, rather than manufacturing. In these sectors, innovations bring only limited efficiencies, so there is little space for high growth: machines have not made lawyers, nurses, or filmmakers 10 times more efficient, and they are unlikely to do so.

Green growth versus degrowth

In the past few years, the mother of all battles has broken out within the group of economic scholars thinking about the future of the planet (admittedly, this is a very small group among economists). On one side are the green growthers. They say that economic growth can be made environmentally sustainable: the economy can "decouple" growth from environmental damage. Lawmakers can find the right policies to reshape the markets and push carbon out of the economy, protect biodiversity, avoid pollution, and prevent further land expansion. At the same time, the economy can keep on growing.

The problem is that empirical evidence for green growth is weak. Many rich countries – Sweden, France, the United Kingdom, and Finland, for example – have impressively reduced emissions within their own borders, but still live beyond the limits of the planet. This is because the goods that their citizens consume are produced in China and other countries where emissions are spiralling upwards.

On the other side of the argument are the degrowthers (or, at the very least, steady-staters). They conclude that degrowth (or, at the very least, no growth) is the only option. This may be a risky economic strategy. Again, the COVID-19 pandemic offered a glimpse of this world: in 2020, greenhouse gas emissions across the globe fell dramatically, and many cities had cleaner air than at any time in many decades. But at what cost? Losing millions of jobs and businesses leads to social unrest and insecurity, the end of innovation, and, potentially, economic collapse. This is a chaotic political environment in which to bring about sustained long-term change. Remember, emissions must fall 7 to 8 per cent per year to remain within the Carbon Law. Moreover, degrowth in rich nations would make it more difficult to bring the hundreds of millions of people out of poverty in poor countries. Who has the right to tell the 800 million people living in extreme poverty, with shocking health and poor education, that economic growth has been abandoned and they should give up hope of a better life?

The planetary boundaries framework has been used by scholars to defend both sides of the green growth/degrowth debate. But the framework is entirely agnostic about the global economy. It only tells us what the contours of the safe operating space are; it does not tell us how the world economy gets there, or how it stays within these boundaries. We are not agnostic, though. We are pragmatic.

First, infinite growth at the expense of the biosphere is impossible. This is not an opinion; it is a statement of fact. Second, economic growth that does not come at the expense of Earth's life support system is possible. Economic growth improves lives, brings security, and slows population growth. Growth is entirely dependent on context. In some places, it is essential to end hunger and poverty. In other places, we should not care one jot about it. In all nations, though, we should move beyond this myopic focus and adopt better metrics for better lives. Japan has struggled with low or zero growth for decades and has now come to terms with it. Countries like New Zealand, Iceland, and Scotland are actively, openly, and courageously exploring policies that put the economy to use to improve the well-being of both people and planet.

So, we should not obsess about economic growth, nor should we demonize it. Our main focus should be on managing markets effectively to support, not undermine, societal goals.

But before we finish on growth, we should recognize one reason why growth is so alluring. Debt has played a big role in creating modern economies. The money borrowed has to be paid back in the future. The big assumption here is that the future will be rosy; after all, the debt can only be paid back if there is some sort of progress or growth. This seems like a good deal for everyone. Debt today can improve future prospects: from student loans to building roads and factories.

The different financial instruments that allow us to borrow from the future operate like a time machine, allowing us to create a social contract with our future selves and future societies who will be paying back the debt. But what if investment today makes the future less stable? In recent years, it has become clear that the social contract is already breaking down; we only have to look at the children striking from school to see it. This is not the deal they signed up for.

Back to school

Now is the time to reject the neoliberalist economic model championed by the Chicago school of economics, which emerged in the 1930s and has dominated the world since the 1980s. It is time to ditch the "winner takes all" economic logic of the past four decades, during which the world's wealth was hoarded by the few rather than used to improve lives and stabilize the planet. Rather, we need to see a return to a more balanced economic pathway in which all economic variables – incomes, outputs, profits, and wages – progress together, benefiting all in society and ensuring stable, political environments. It is time to go back to school.

The foundation of the new school of economic thinking, let's call it the Stockholm school, is the three "R"s: resilience, regeneration, and recirculation. These largely cover the parts of the economy that relate to the use of physical "stuff". In the school, we also have the KIDSS: the knowledge economy, the information economy, the digital economy, the service economy (including education and health care), and the sharing economy. The KIDSS can grow. Taken together, they are the building blocks of a new economic model that drives a resilient well-being economy within planetary boundaries.

The three "R"s and the KIDSS

Here are the rules of the new "doughnut" economic playing field.

Rule #1. Resilient people make for a resilient economy in times of transformation and turbulence.

COVID-19 showed that an economy is nothing without people. The most essential workers during the crisis were often relatively low-paid: care workers, delivery drivers, cashiers, warehouse workers, paramedics, nurses, teachers, and journalists. The tens of millions of people who lost their jobs face economic hardships that might last decades. The young students graduating into a global recession may never in their careers have the level of opportunity afforded those who graduate in good times.

We can expect more economic shocks in the coming decades, from infectious diseases, automation (and eventually AI), cyber attacks, banking failures, climate change and other environmental calamities, and, of course, political failures. These shocks will be characterized by speed, scale, connectivity, and surprise. Can we even think about building economies resilient to these threats? Yes. But this means protecting people.

First, we need to protect people's health by investing in effective safety nets, such as universal health care. This is not some "nice to have" if the economy can afford it; it is an essential public good for a resilient economy in the Anthropocene.

In times of transformation, we also need to protect people's job security. Even without unexpected shocks, economies must transform; the fossil fuel industry must contract. There will be turbulence. Many jobs will be created; others will disappear. With the right support in place, this can be an opportunity for employees and employers to advance. Providing free (or subsidized) education at all stages of life will allow people to change jobs with less friction, and governments should also seriously consider ideas like universal basic income. This is an investment in resilience. And cross-pollinating industries with a steady stream of new people from other sectors is a surefire catalyst for innovation.

Finally, protecting people means redistributing wealth fairly. The evidence is unequivocal that greater economic equality reduces crime, obesity, and drug use and also improves collective decision-making for the common good. This is not an economic fantasy world. This is the Nordic model. If it is not Denmark or Sweden, it is Norway or Finland that regularly tops international league tables for health, economic equality, gender equality, well-being, education, tolerance, democracy, and trust. So, economies need to shift away from thinking about efficiencies and towards building resilience.

Rule #2. Regenerate natural resources.
An economy within planetary boundaries stores carbon rather than emits it, enhances biodiversity rather than destroys it, and protects soils and waters rather than pollutes them.

The foundation of the new economy is regeneration. Nature is spectacularly good at living within its means, and there is much we can learn from it. Evolution ensures multiple organisms contribute to regeneration and recycling of all materials. Rich ecological diversity renews itself continually. From fishing grounds to farms, from deforestation to mining, industries must be channelled to support biosphere stewardship and enhance the resilience of our living systems.

This regenerative economy is already taking off. About 24 per cent of electricity currently comes from renewable sources and, as we discussed in Chapter 11, close to one-third of farms worldwide are now practising some form of sustainable farming. Through initiatives such as the Seafood Business for Ocean

Stewardship (SeaBOS), we are seeing the world's largest seafood companies come together to chart out a sustainable path for our ocean. And recent research shows that if 90 per cent of new buildings were built from wood over the next 30 years, this would lock 20 billion tonnes (22 billion tons) of carbon out of the atmosphere. This is well over one year's worth of current emissions.

Rule #3. Recirculate everything.

If people are at the heart of the economy and a regenerated biosphere is the foundation, then the circular economy is the flywheel creating unstoppable momentum. At the moment, every person on Earth consumes an average of 13 tonnes (14 tons) of "stuff" every year. From minerals to fossil fuels, from crops to trees, the world eats through a staggering 100 billion tonnes (110 billion tons) of material. Of course, most of this consumption is happening in rich nations. We use most of the material for housing, factories, roads, and other buildings. But transport, food waste, and consumer goods are big ticket items, too. About one-third of all these materials are dumped. Economically, this is a phenomenal waste of money. The good news is that close to 10 per cent of the global economy is now circular. The potential is vast. In Europe, a more circular economy could cut emissions from heavy industry by 56 per cent by 2050.

Some of the world's most valuable companies have already committed to circular business models. IKEA plans to be climate neutral by 2030. The company announced that all products will use only renewable and recyclable materials and will be developed to be repurposed, repaired, reused, resold, and recycled. Similarly, fashion giant H&M has pledged to use 100 per cent recycled or sustainable materials by 2030 and to become fully climate positive by 2040. These economic models are moving rapidly into the mainstream, as companies realize that many of the most used materials are fully recyclable: steel, aluminium, plastic, paper and cardboard, glass, and food waste. Even concrete, a spectacularly energy-intensive material to produce, can be recycled endlessly.

Rules 2 and 3 are based on the foundational principle that we must now end economic growth based on unsustainable extraction of resources. So where will new growth come from?

Rule #4. Let the KIDSS grow.

The knowledge, information, digital, service, and sharing economies (KIDSS) are free to grow. They are tightly interlinked and will be the source of innovation and creativity this century. A knowledge and information revolution will provide the foundation for the economic growth needed to allow the future population to live good lives on Earth. This is clearly happening already: many developed economies are now based on services, knowledge, and information rather than manufacturing.

Central to the KIDSS economy is sharing. Buildings, tools, and vehicles often sit idle for much of the time. Digitalization can end the headache of connecting people who have cars, tools, or space with those who want to use them temporarily. Shifting to a service-based business model, where resources are unlocked for others to use, can boost profits and reduce emissions simultaneously. Novel business models are popping up in surprising places. Take farming: farmers have traditionally bought fertilizer and spread it on their crops. But this can be wasteful: a farmer may lack the information and knowledge about the best time, dose, and location for maximum yield. Now, farmers can buy monthly fertilizer services, where companies assess land, weather, and growing conditions to optimize fertilizer application, thus reducing waste and water pollution. Think of it as Netflix for farmers. This model can apply to many other aspects of farming and, indeed, to other industries.

But the KIDSS economy risks transmuting into the surveillance economy we introduced in Chapter 15, characterized by data extraction to exploit consumers. Without good governance, this could accelerate unsustainable behaviour and destabilize democracies.

Earthshot economics

So, how do we turn all of this into an economic plan of action? More specifically, a plan that reinforces social progress and nature-positive economic activity. The two solutions are startlingly simple: reshape the markets and return to long-term economic planning.

First, we need to reshape the markets to make it unprofitable to produce things that work to destabilize the planet and profitable to produce things that help people and planet. Currently, few countries put a price on carbon (for example, a tax or other financial

PATHWAYS TO STABILIZE EARTH

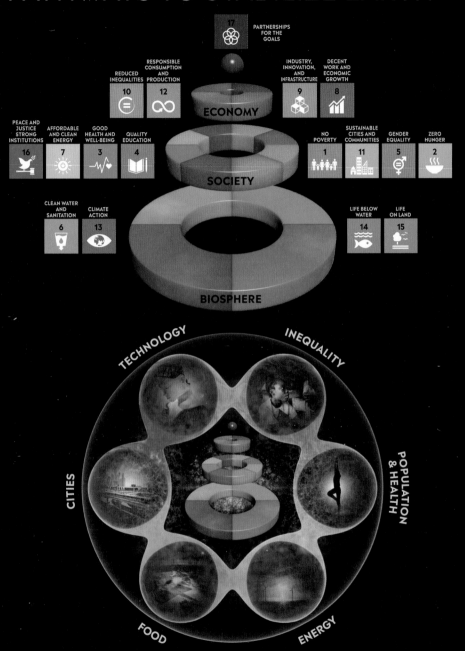

17 PARTNERSHIPS FOR THE GOALS

RESPONSIBLE CONSUMPTION AND PRODUCTION
REDUCED INEQUALITIES
10   **12**

INDUSTRY, INNOVATION, AND INFRASTRUCTURE
DECENT WORK AND ECONOMIC GROWTH
9   **8**

ECONOMY

PEACE AND JUSTICE STRONG INSTITUTIONS
AFFORDABLE AND CLEAN ENERGY
GOOD HEALTH AND WELL-BEING
QUALITY EDUCATION
16   **7**   **3**   **4**

NO POVERTY
SUSTAINABLE CITIES AND COMMUNITIES
GENDER EQUALITY
ZERO HUNGER
1   **11**   **5**   **2**

SOCIETY

CLEAN WATER AND SANITATION
CLIMATE ACTION
6   **13**

LIFE BELOW WATER
LIFE ON LAND
14   **15**

BIOSPHERE

TECHNOLOGY   INEQUALITY

CITIES   POPULATION & HEALTH

FOOD   ENERGY

In 2015, the United Nations member states agreed 17 Sustainable Development Goals (top), ranging from ending poverty to improving human health. But the world will only achieve long-term sustainability if Earth's natural systems are resilient and functioning effectively.

The Earthshot mission is based around six system transformations (bottom) to stabilize our planet and meet the Sustainable Development Goals. There is growing scientific consensus that if we get these six transformations right, 10 billion people will be able to live prosperous, healthy lives on a stable planet.

D1

LOOK BACK AT WHAT WE HAVE ACHIEVED

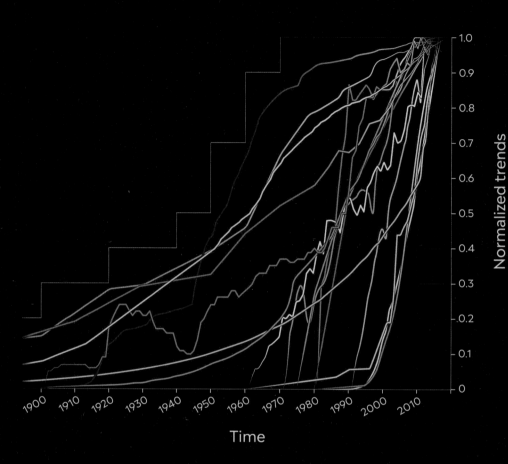

Normalized trends

Time

Through science, democracy, and cooperation, humanity has achieved the impossible in the past 50 years. Universal literacy is within striking distance. Our knowledge of the planet is accelerating. We are protecting more land and ocean. The majority of people on Earth have access to the Internet or mobile phones. In addition, the end of extreme poverty is within sight, and global life expectancy at birth is now 72 years.

- Literacy
- People not in extreme poverty
- Life expectancy
- Scientific articles
- Democracies
- Women's vote
- Protected areas
- Harvest
- Monitored species
- Girls in school
- Child cancer survival
- Access to water
- Immunization
- Mobile phones
- Access to Internet
- Electricity coverage

THE GLOBAL SAFETY NET

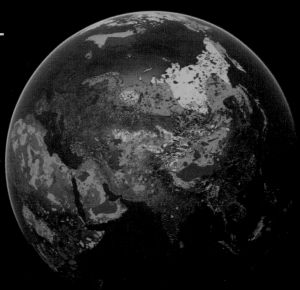

■ **Current global protected areas (15%)***

▦ Additional unprotected areas needed to conserve sites of **Species rarity (2.3%)**

▦ Additional unprotected areas needed to conserve sites of **Distinct species assemblages (6%)**

▦ Additional unprotected areas needed to conserve sites of **Rare phenomena (6.3%)**

▦ Additional unprotected areas needed to conserve sites of **Intactness (16%)**

▦ Additional **Climate stabilization areas (4.7%)**

∿ **Wildlife and climate corridors** (to connect intact habitats)

* Including polygons selected for Species rarity, Distinctness, Rare phenomena, and Intactness

In 2020, for the first time, scientists calculated and visualized what it will take to protect half of the land on Earth. The Global Safety Net maps show how expanding nature conservation addresses both climate change and biodiversity loss. The research identifies 50 per cent of the terrestrial realm that, if conserved, would reverse further biodiversity loss, prevent carbon dioxide emissions from land conversion, and enhance natural carbon removal. This framework shows that, beyond the 15.1 per cent land area currently protected, 35.3 per cent of land area is needed to conserve additional sites of particular importance for biodiversity and to stabilize the climate. Indigenous lands overlap extensively with the Global Safety Net. Conserving these sensitive areas could support public health by reducing the potential for zoonotic diseases such as COVID-19 to emerge in the future.

THE RISE IN COMPLEXITY: MAJOR RUPTURES IN THE EVOLUTION OF EARTH

Earth 5.0
WISE

Earth 4.0
COMPLEX

We are here

Earth 3.0
SOLAR

Earth 2.0
LIFE

Earth 1.0
PRIMEVAL

| 4.6–3.5 billion years ago | 3.5–2.4 billion years ago | 2.4 billion–600 million years ago | 600–0 million years ago | Future |

Earth's history is punctuated by three evolutionary events that changed the course of the planet: the emergence of life (Earth 2.0), photosynthesis with oxygen (Earth 3.0), and multicellular complex life (Earth 4.0). Each step led to a leap in complexity and with it a greater ability to process information about the immediate environment. Processing information is a life or death matter for living organisms. With the arrival of *Homo sapiens*, this level of complexity has jumped again. One species is now able to process information relating to the entire Earth system. Does this mark a rupture as profound as the arrival of multicellular life or photosynthesis? Possibly.

instrument), but everywhere has a tax on labour, which contributes to driving rapid automation. This seems the wrong way around. We want to tax things we don't want – pollution – and lower taxes on things we want more of – quality employment. Redressing the balance so that taxes rise on carbon and fall on labour can encourage changes in behaviour of both consumers and producers towards lower emissions and greater employment.

Second, long-term planning or cathedral thinking. In the past, long-term planning meant having lots of children in the hope that some would survive to look after us in our old age. But some societies had a bigger vision, and the deep pockets to match. The Egyptians, Maya, and medieval religious leaders who built pyramids and grand cathedrals set foundations in place, knowing that neither they nor their children would see the final marvel. They had a higher goal. We need more of this cathedral thinking, and the main institutions that can deliver on this are governments.

In order to operate within planetary boundaries, the world needs more than just renewable energy. We need to redesign electricity grids and construct high-speed railways, tunnels, bridges, and highways, fully connected to the new grid. We must build more electricity storage, retrofit buildings on a grand scale, and convert the global shipping fleet to zero emissions in a generation. But we need to think transformatively: we are not talking about adding 1 metre (3 feet) to a levee to protect from flooding, a minor engineering feat. We are talking about building sponge cities, designed to absorb water (and other shocks). This investment is not only critical to save the planet; it is the basis of a new social contract with our future selves and the investment opportunity of our lifetime. These are the projects that will create the essential economic optimism that our future will be better than our past.

Before we finish with the global economy, let's take a glimpse at the dark side.

What is the worst that could happen?
Imagine that governments do nothing, or very little, to reshape markets. Imagine the finance sector carries on regardless. Imagine the extremes – strong hurricanes, intense heat, megafires, and floods – become more extreme.

"Will cities be able to afford their infrastructure needs? What will happen to the 30-year mortgage – a key building block of finance – if lenders can't estimate the impact of climate risk over such a long timeline? What if there is no viable market for flood or fire insurance in impacted areas? What happens to inflation, and in turn interest rates, if the cost of food climbs due to drought and flooding? How can we model economic growth if emerging markets see their productivity decline due to extreme heat and other climate impacts?" This is not scaremongering from environmental groups. The speaker is Larry Fink: the CEO and chairman of the world's largest asset management company, BlackRock. He is directly addressing the CEOs of the companies that BlackRock backs.

Without deep transformation, the finance sector is crowdsourcing catastrophe. There are three probable scenarios for how market confidence might evolve in the 2020s: Carbon shock, Carbon shock plus, and Stable transformation.

Carbon shock
In the first scenario, the markets sense politicians are weak and do not have the appetite for strong climate policies. Demand for fossil fuels continues to rise steadily, and prices remain relatively stable. But regardless of political will, or lack of it, the global economy has already reached a point of no return. The clean technological revolution is an unstoppable juggernaut. The fossil fuel era is over. Demand for fossil fuels slumps in the late 2020s, and the carbon bubble bursts, leaving USD 1 trillion in stranded assets as investors are left with a pile of worthless pipelines and refineries.

Carbon shock plus
The second scenario begins similarly to Carbon shock. Politicians fail to translate their rhetoric on the threat posed by climate change into adequate action, sending a signal to markets to continue as normal. Then, in the mid 2020s, they have a change of heart. Perhaps public pressure becomes so unbearable that inaction becomes untenable, or maybe improvements in climate models show that Earth is destined to warm even faster. With that, our remaining carbon budget would disappear. Another possibility would be a major environmental catastrophe in the 2020s.

Maybe the jetstream stalls over the world's breadbaskets (regions that are particularly suitable for growing wheat and maize), bringing global drought and severe food shortages. Or a dramatic escalation of fire spreads around the world, as forests and peatland dry out and become more prone to combustion. Or parts of the Antarctic ice sheet crack and disintegrate, setting off alarm bells across the planet.

The result is that lawmakers will be forced to take draconian action. This is likely to shift market sentiment forcibly away from fossil fuels, thus setting off a chaotic destabilizing stampede. By one estimate, this could lead to a carbon bubble closer to USD 4 trillion. Inevitably, there would be winners and losers. The losers would be countries who clung to fossil fuels for too long, such as the United States, Canada, and the Middle East. The winners would be China and Europe, who may have a relatively limited financial exposure to fossil fuels and will benefit from energy independence.

Stable transformation

In this scenario, politicians send strong, immediate, unequivocal signals that they will redirect the markets towards Earth system stabilization. They announce ambitious phase-outs of internal combustion engines and set targets to reach zero emissions, zero deforestation, and zero biodiversity loss. And they put the policies in place (see Chapter 17) to drive this. The markets respond by reallocating investments towards these long-term goals and away from fossil fuel resources that will increasingly dry up as governments ratchet down. This managed transition scenario buys governments time to invest in regions and workers negatively affected by the redirection of capital.

The three probable scenarios highlight the difference between an ordered exit from fossil fuels and an economic calamity, like COVID-19 and the 2008 financial crisis. We would prefer the ordered exit. As former United Nations climate chief Christiana Figueres said so eloquently, "We will move to a low-carbon world because nature will force us, or because policy will guide us. If we wait until nature forces us, the cost will be astronomical."

EARTHSHOT POLITICS AND POLICIES

Fixing markets isn't enough. We have to actively shape and create them and tilt the playing field in the direction of the growth we want.

MARIANA MAZZUCATO
ECONOMIST, 2016

The COVID-19 pandemic changed everything. It exposed not only a deep fragility in our systems for governing, but also a profound solidarity and common humanity. The crisis allowed us to think the unthinkable. Governments put human lives before the economy, and trillions of dollars were found in back pockets to shore up economies.

Undoubtedly, COVID-19 was the biggest global shock since the Second World War. The war led to a disruptive reshaping of political systems to create more equality and to engender greater cooperation internationally. This ushered in decades of peace, prosperity, and progress based on an efficient global economy. Will the pandemic recovery prompt an effective reorganization of international politics to support a resilient global economy operating within planetary boundaries? It simply must.

Governments have found themselves in an unexpectedly powerful position: they might reasonably place some conditions on bailouts. Why? Because they will be paid for by future taxpayers. And who are these taxpayers? Well, our children, the young generation who, even before the pandemic, made it quite clear that the climate crisis is pulling the rug from under their feet. Now, those in power today are asking future generations to pay invoices for climate extremes and COVID-19 debts. One can easily see that this is the script for deep indignation. Political failure at this moment is unforgivable. We have reached the end of the line.

The Moonshot of the 1960s was testament to the power of a political vision, bringing a nation together around a singular mission. It was a uniquely coordinated effort between government, research, and industry, and it cost the United States 2.5 per cent of its GDP. The sheer scale of the vision truly inspired the world, which is precisely why we need an Earthshot right now. Imagine if 2.5 per cent of global GDP were invested in stabilizing the planet. Globally, this would mean devoting almost USD 2 trillion per year to save our planet and to lift our societies to the next level of technology, health, and well-being.

The Moonshot was not the first time industrial nations channelled the power of their economies towards a mission-driven challenge. Two decades earlier, the Second World War forced engineers and scientists to focus on a new task – supporting the war effort – and factories shifted to the production of aircraft, tanks, and guns. In the same way, during the 2020 pandemic, universities, companies, and industries dropped what they were doing to unravel the genetic codes of the virus, develop and scale testing systems, and invent vaccines to protect the global population for the common good.

Crucially, a mission-driven vision to save our planet is drawing attention. In 2018, Italian-American economist Mariana Mazzucato published a report defining what type of projects are needed to fuel innovation. Like the Moon landings, they must be bold and must inspire citizens to get behind them. The projects need a clear target and deadline, and must allow experimentation for ideas to bubble up. Finally, they must pull in diverse groups of researchers. The Green Deals that are popping up around the globe (see Chapter 12) are certainly in this vein, at least in part.

The Earthshot mission to stabilize Earth will succeed if we make the case for it as simple and as attractive as possible. A big part of the latter is to make it profitable. At the same time, we must go beyond a singular focus on energy and emissions to embrace an agricultural revolution and enhance the resilience of all the remaining natural ecosystems, from rainforests to peatlands.

Yes, planting a trillion trees is a good target that rallies people, but this must be accompanied by a far deeper understanding of ecological resilience, as our goal is to preserve and nurture thriving natural ecosystems. The long-term target needs to be crystal clear: cut greenhouse gas emissions by half, zero loss of species, and zero depletion of natural ecosystems by 2030 at the latest (against a baseline of 2020). We require ambitious projects that build clean, green railways, roads, cities, and power while redistributing wealth. The European Union, New Zealand, Costa Rica, and the United Kingdom are moving in this direction, showing the world that it is possible. This action will require novel ways to fund long-term infrastructure projects. Governments should create new partnerships with private investors to build the future. There is no shortage of cash in the markets, only a lack of visionary projects to invest in.

Policies for a stable planet

There are four political levers to pull to reshape the markets in order to drive Earth-centric and people-centric economic development. These four proposals are perfectly aligned with the latest science and economics of global sustainability.

1. Immediately pass legislation to achieve net zero greenhouse gas emissions by 2050 at the latest and net positive nature by 2030.

Two countries have already achieved one of these goals and become carbon neutral (Suriname and Bhutan), five countries have it in law (Denmark, France, New Zealand, the United Kingdom, and Sweden), and dozens of countries are discussing it. At the centre of the European Union's Green Deal is a target of net zero carbon emissions by 2050. But the richest nations have a historic responsibility to reach net zero earlier, for example by setting a hard target of net zero by 2040.

For nature – biodiversity and critical global commons such as forests and wetlands – governments must urgently end all loss. By 2030, the process of recovery and building resilience must have started. The goal is that by 2050 we will have full recovery, restoration, and regeneration. At this point, we will have achieved sufficiently functioning ecosystems to support future generations.

2. End all investments in new fossil fuels.

This means pipelines, refineries, coal mines, and coal-fired power plants – everything. This is the beginning of the end of the fossil fuel era. Right now, in the midst of this planetary emergency, countries around the world are building, or have plans to build, a further 1,200 coal plants. Existing power plants will emit a total of 660 billion tonnes (727 billion tons) of carbon dioxide if they run to the end of their natural lives – this amount is greater than (roughly twice) our allowable carbon budget if we are to have any chance of holding temperature increases to 1.5°C (2.7°F). In short, coal plants, if allowed to expand as planned, will on their own gobble up the remaining global carbon budget. Proposed power plants would eat another 190 billion tonnes (209 billion tons). Further expansion is crazy, but we also need to be pragmatic. Inevitably, some polluting plants will be built, and their emissions will push the world past its allocated budget. This is why, like it or not, major investment in carbon capture and storage – burying carbon dioxide deep underground, for example – will be necessary.

3. Withdraw all subsidies that promote fossil fuel use, biodiversity loss, and deforestation.

We know that governments are still paying more than USD 500 billion a year in direct subsidies to the fossil fuel industry, and 10 times this amount when you include the harm to health, the environment, and economies from fossil fuels. The world's largest economies have said they will phase out these subsidies, but nothing is happening. Subsidies for farmers also often work against efforts to protect biodiversity and carbon sinks. We need to change agricultural subsidies so that they support efforts to build carbon sinks, restore wildlife, and end farm expansion into virgin lands.

4. Put a price on carbon.

About 80 per cent of the world's emissions of carbon dioxide are emitted entirely free of charge. Even the most liberal market economists agree that it is a serious malfunction of markets to have unpriced harmful so-called "externalities". Destroying air quality and climate stability for free is the world's most serious market failure, and everyone knows it. The problem is that it has served many economies (read, rich economies) very well over the past 100 years. But the atmosphere is global, a shared resource we all depend upon. We will fail to stabilize the planet if a carbon price is not applied across all sectors. This means a global price on carbon has to be adopted by all nations.

There is a general consensus among economic researchers analysing climate policies that countries need to set a price of at least USD 50 per tonne (1.1 tons) of carbon dioxide applied to all sectors, from food and transport to heating. No exceptions. Sweden now has a price on carbon that is equivalent to about USD 120 per tonne. Canada has implemented a carbon tax, in which all the money raised is paid back to taxpayers to make it fairer: if you live a low-carbon lifestyle, you are rewarded accordingly. The price on carbon does not have to be imposed top-down and be designed universally. It can emerge in various shapes and forms across different economic regions.

When the network of nations, companies, and cities that apply a price on carbon becomes large enough, and connected enough, this will eventually tip over the global system, thereby mainstreaming a price on carbon throughout the entire world. This, in turn, will be the true kiss of death for the fossil fuel-driven world economy. Some of the revenues from a price on carbon should be funnelled into an international fund to invest in carbon capture. Governments will receive large inputs to their state coffers from a price on carbon. Much of this can and should be returned as social dividends to compensate low-income households.

Politics for a stable planet

If the policies we need to stabilize the planet have so many benefits for societies, why is there such resistance to change? Well, the science is demanding a profound upheaval to the foundation of the economy: energy. This cannot be taken lightly. The fossil fuel companies have become the world's most powerful industry, wielding formidable political might. On top of that, it is future generations that will have to deal with the greatest impacts, not today's politicians, so it is easier to kick the can down the road.

The good news is that an energy revolution is coming anyway. It is now unstoppable. The four policies outlined earlier will lock it in and accelerate it. As with all previous energy revolutions, the new one will create more jobs, as many as 40 million. All major economies are built on renewal and rebirth. This is nothing new. The history of capitalism is littered with companies and industries that have collapsed and died as innovation made them obsolete. This time we need a just transition. We must protect workers through major investment in retraining and education.

In the past three decades, globalization, digitalization, and automation have destroyed traditional manufacturing jobs and industries as companies in wealthy nations have relocated to China, where labour is cheaper. However, some countries, such as Sweden, have invested in resilience to economic change through a culture of life-long learning. The Swedish government supports companies and society by investing in retraining and by providing strong economic safety nets for people. This is the right approach. Spain is providing USD 250 million to support the closure of its coal mines, with investment in the workforce and new industries. Germany has followed suit: USD 32 billion has been put on the table as part of the climate package adopted at the end of 2019 to invest in the transition away from coal. In the United States, however, workers are largely abandoned when companies shutter up. When an industry closes in a US town, people are expected to pack up and move elsewhere to find similar work. In reality, the inertia of family ties and the uncertainty of moving mean that this is unlikely to happen. The gales of creative destruction sweep in, bringing poverty, misery, and stigmatization. Some towns never recover. This is a failure of capitalism that can be avoided with long-term planning.

There is no doubt that we are living in a politically challenging time: the current climate is an emotional roller-coaster of highs and lows in international politics. There was genuine euphoria in 2015 when the United Nations navigated all national leaders towards agreeing the Sustainable Development Goals – a vision for our common future – followed three months later by a global deal on climate. In 2016, however, things began to unravel with the surprising results of the Brexit referendum and the US presidential election, followed by the Brazilian election in 2018, and so on.

The door opened to demagogues who have no interest in complexity and who reassure voters that simplistic solutions – halting migration – will solve everything. They won't. For planetary stewardship and the Earthshot mission to succeed, we need to rebuild trust in democratic institutions and global cooperation. This is our priority. Mainstream political parties must extend the olive branch, bury the hatchet, and find common ground.

This means acknowledging the legitimacy of political opponents and re-establishing a shared understanding of facts, evidence, and science. The Earthshot cannot emerge without these foundations. More efforts must go into reducing polarization. In countries without proportional representation, this could involve creating unity governments, in which the winning party in an election invites members of opposition parties to serve alongside them in government.

The most important mechanism for establishing trust, though, is reducing inequality. More equal countries tend to have greater trust in governments, which makes collective decision-making for long-term goals easier. So, in addition to an international price on carbon, a global wealth tax is a very good idea. This is because unrestrained growth in inequality will rip this planet in two.

Conversely, a fairer redistribution of wealth will rebuild trust and help instil a world view based around solidarity and a common human identity. A progressive global wealth tax would do more good for more people for longer than all the philanthropy in the world today. It needs to be global, though, because otherwise the rich will find ways to bury their treasure in some remote haven.[24] Many economists, including Thomas Piketty and Joseph Stiglitz, advocate such a tax.

On top of a wealth tax, we also suggest a global corporation tax. Multibillion-dollar companies pay a pittance in tax as countries vie to offer them the lowest tax rates. It is a race to the bottom in which everyone loses but the company. We simply need to set a minimum global corporation tax rate. Everyone wins. Indeed, the conservative

---

24 We wonder, what is the tax status of the Moon and Mars?

Organisation for Economic Co-operation and Development, the bastion of globalization, now endorses a global corporation tax.

Seventy-five years ago, the United Nations emerged from the rubble of war. This was a historic moment for international politics, which also saw the birth of the World Bank, the precursor to the World Trade Organization, and the International Monetary Fund. These global institutions were not designed to deal with the speed, scale, and surprise of the Anthropocene, though. So, perhaps now is the time to reopen discussions about adopting more democratic principles at the global level – everyone might one day vote in a world parliament. That may seem a long way off, but the United Nations appeared amid a backdrop of chaos and war. Now, the global pandemic is likely to cause the biggest upheaval since that time. It is a moment for new ideas and new thinking to emerge.

Trust: share of people agreeing with the statement "Most people can be trusted"
Income equality is an important factor in establishing trust in both society and government. Trust is essential for collective decision-making. (Representative number of countries shown)

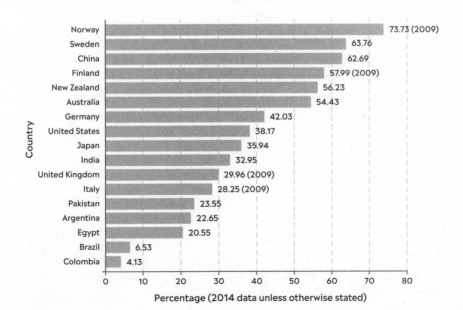

THE ROARING 2020S: FOUR TIPPING POINTS ARE CONVERGING

Look at the world around you.
It may seem like an immovable, implacable place.
It is not. With the slightest push – in just the right
place – it can be tipped.

MALCOLM GLADWELL
*THE TIPPING POINT: HOW LITTLE THINGS CAN
MAKE A BIG DIFFERENCE, 2000*

As 2019 drew to a close, an unknown virus leapt from a wild animal, perhaps via a domestic animal, into its first human victim. In one way, this might be viewed as the famous "butterfly effect" in action: a seemingly insignificant event that can change the course of history. But it is also an example of exponential scaling. We now know that doubling again and again reaches a dizzying scale rather rapidly. The epidemic has shown us unequivocally that small things can have a very big effect in the connected world of the Anthropocene. And change can happen very quickly. Ultimately, the pandemic will, in all likelihood, lead to a shift in awareness about what it means to be human in a deeply connected and vulnerable world.

We are cautiously optimistic about the future. Our optimism is based on the simple observation that four colossal forces are aligning to pull Earth into a future within planetary boundaries. We believe that the world is reaching or has crossed beneficial tipping points in four key areas: social, political, economic, and technological. Each of these tipping points has been gaining momentum year by year, and sometimes even week by week. This growth has been exponential. The thing with exponential change is that it is deceptively slow at the start, until you reach the knee of the curve, and then things take off. Currently, the world is at the knee of the curve. The next decade is on course for the fastest economic and social transformation in history. Welcome to the roaring 2020s.

Major leaps forward in society are driven by disruptions from social movements, government policies, market confidence, new technologies, and science, or some combination of these.

The end of apartheid and child labour or the arrival of women's rights and civil rights followed a pattern of growing social momentum, until a tipping point forced open the floodgates. Or take smoking in public places. Once one government took a blind leap into the unknown and banned the activity, this created a groundswell of support. People flipped from seeing only downsides – pubs closing, restrictions on god-given freedoms – to finding a bewildering array of benefits, from fresh air in restaurants to reduced fire risks, fewer heart attacks, and changing attitudes of young people to smoking. A different type of tipping point occurs in economics: if the price of a new technology falls below that of the old, and the technology is constantly improving, this creates an irresistible combination that sets up a self-sustaining and amplifying feedback loop. Electric cars will triumph over the internal combustion engine, not because they are the "right" way forward but because they outcompete on price and performance. Another way to bring about massive transformation is through new ideas and innovations – often originating in academia or the technology sector, but in reality they can come from anywhere – which open up new markets, new ways of thinking, new modes of living.

In this chapter, we will explore each of the four tipping points.

First, we will discuss the explosive social phenomena of the Fridays For Future school strikes, Extinction Rebellion's direct action, and the growing public awareness that we are experiencing a planetary emergency. Second, we will look at the abrupt rise of Green Deals in politics and the changing political landscape. We cannot say yet how the COVID-19 global crisis will finally resolve, but it has already proved that massive and swift political action is possible. And third, we will address economic tipping points, as the prices of clean, green solutions drop below those of polluting products and carbon bubbles begin to grow. A strong feedback loop is building; the exit from fossil fuels could be quick and beneficial overall, but brutal for a few.

We will finish with the technological revolution, which will, over the next decade, transform how we work, how we live, and how we consume and commute, as well as how we look after our health and develop our minds.

1. The social tipping point

In September 2019, the school strikes for the climate attracted between 6 and 8 million people, according to various estimates. Strikes occurred in 4,500 locations in 150 countries around the world. Regardless of the precise figure, this was by far the largest climate strike in history. In fact, it was one of the largest single demonstrations in modern history, on a par with the student risings in Paris 1968, the protests against the Vietnam War in the 1960s and 1970s, and the rallies against the 2003 invasion of Iraq.

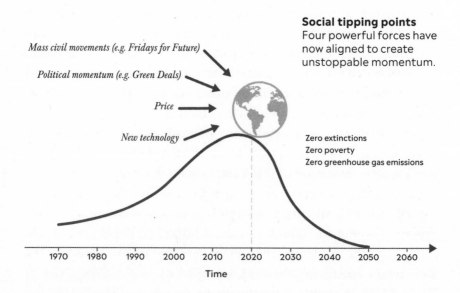

Up to that point, the school strikes had been on a phenomenal exponential journey. The previous 13 months had demonstrated the power of the network effect in the modern world. It all started in August 2018, when a Swedish schoolchild, Greta Thunberg, lugged her sign *"Skolstrejk för klimatet"* (School strike for the climate) down to the parliament building in Stockholm. This image of a lone girl – she had not been able to convince her classmates to join her – making a stand caught the attention of the Swedish media. After an explosion of support for her on social networks, the global media joined the party. Her sign has since become iconic.

In his essay in *Nature* in 2011 on the role of social networks in fuelling the Arab Spring, Philip Ball argued that networks "enabled a random event to trigger a major one". The self-immolation of a street vendor, Mohamed Bouazizi, in Sidi Bouzid, in protest at harsh treatment by officials, tipped Tunisia into a chaotic revolt against the ruling elites. Ball noted, "Three months earlier, a similar thing had happened in the city of Monastir – but few people knew about it because it was not publicized on Facebook." The new media landscape and our phenomenally networked society have changed the dynamics of what is possible.

Future historians will look at the moment when schoolchildren abandoned their studies to strike for their future – the Fridays For Future movement – as evidence of a deep societal rift opening up. Children are now calling out older generations. Incredibly, to get their facts in order, children are bypassing the mainstream media and going straight to the scientific literature in order to fathom what is going on. They have discovered that a slow departure from fossil fuels over a century will disastrously destabilize the planet, and that future generations will not stand a hope in hell of restabilizing Earth unless we embark on a deep global transformation immediately. This realization has rightly prompted anger. Our children are shocked and dismayed. They have been sold a lie. Their action is perhaps the first real sign that planetary stewardship has truly gone mainstream.

Recent research confirms that children can influence their parents' concerns around the climate crisis. Children are able to do this because we do not perceive their ideas as coming from political ideology. Furthermore, parents care what their children think on a range of issues. The researchers noted that the largest change was seen among conservative parents, particularly dads. And daughters are especially influential.

At her second appearance at the World Economic Forum in Davos in 2019, Greta complained to business leaders that they had ignored her message. She said the world had made no progress at all in the previous 12 months. But there has been a perceptible shift. Even before the pandemic and its impact on travel, the Swedish phrase *flygskam* or "flight shame" had taken off, not only in Sweden but also across Europe, resulting in a real drop in the number of flights from Swedish airports. Train use is increasing, too. The national Swedish rail company is now planning to expand its sleeper train coverage to Paris, Brussels, and London.

The Fridays For Future movement has reshaped the conversation on climate, riling Donald Trump and Jair Bolsonaro in the process. This has been dubbed the "Greta effect". Even newspapers are getting in on the act. The editorial board of the *Financial Times* wants business travellers to eschew business-class flights and opt for night trains instead. They list multiple benefits. Travelling by train allows people to save money on hotels, avoid miserable early morning flights, and skip security queues and baggage carousels. Pricing aircraft emissions fairly, "rather than throwing subsidies at operators", would accelerate the transition.

In 2019, the head of OPEC, a powerful group of countries that control much of the world's oil production, called climate strikers the "greatest threat" to his industry. Greta responded on Twitter, "Thank you! Our biggest compliment yet!"

The school strikes have become the public face of a much larger movement. Extinction Rebellion (or XR, as it is known) was also established in 2018, but this time in the United Kingdom. In April 2019, XR brought traffic to a standstill in parts of London, using direct but peaceful action. Protests later spread to New York, Berlin, Paris, and beyond. XR has three clear demands. First, that the UK government declares a climate emergency. Second, that the United Kingdom legally commits to reducing carbon emissions to net zero by 2025. Third, that a citizens' assembly is formed to oversee the changes. Two of these goals have now been achieved to some extent. The United Kingdom has declared a climate emergency, and a citizens' assembly has been established, but the carbon emissions target is unlikely to be met.

For more than a century, non-violent resistance campaigns have been more than twice as effective as their violent counterparts in achieving their goals. Protests, boycotts, civil disobedience, and other forms of non-violent non-cooperation attract impressive support from citizens. Young people are most likely to agree with XR's approach of direct but peaceful action. Indeed, a survey of 3,000 people found that 47 per cent of 18 to 24-year-olds supported the disruption of traffic and public transport to highlight XR's aims.

It is clear that social movements and networks are building around the planetary emergency. New world views and norms are emerging, sometimes in the oddest of places. Take climate sceptics, for example. For decades, television broadcaster and newspaper columnist Jeremy Clarkson used his media platforms, such as the BBC series *Top Gear* and his column in the tabloid newspaper *The Sun*, to ridicule climate science. In 2019, when confronted with the impacts of climate change while filming in Cambodia and Vietnam, he confessed that he now found climate change "genuinely alarming".

The story is similar among business leaders. We are seeing a monumental shift in CEOs engaging more seriously in climate change and sustainability. In September 2019, we sat down for breakfast with Henrik Henriksson and Åsa Pettersson, the CEO and head of corporate affairs at Scania, the Swedish trucking company. We began talking, appropriately enough, about Greta's influence, and discovered that the school strikes have had a profound impact on Henriksson. On the day of the strikes, Henriksson authorized Scania employees around the world to down tools for at least one hour. He encouraged the workers to use the time to attend courses on climate change – Scania's own climate strike. Seeing solidarity between a multinational corporation and schoolchildren, and an acceptance of a new responsibility for the planet, is remarkable. It indicates that concern about the impact of climate change is becoming widespread. Some 68 per cent of Australians believe climate change is a serious threat to their way of life. In the United States, nearly 60 per cent of Americans are now either "alarmed" or "concerned" about global warming. This figure has tripled in the past five years. Ahead of the World Economic

Forum in 2020, a poll showed that almost one-quarter of CEOs are now "extremely concerned" about climate-related issues.

These statistics provide a strong case that we are crossing a social tipping point. According to recent research, minority groups only need to reach 21 to 25 per cent of a population to cross a tipping point and drive sweeping, unstoppable changes in social conventions. Here is a simple scenario to demonstrate this. Imagine a family of four eating dinner at home each evening. One day, the daughter announces she is going to try a vegetarian diet. After a lot of discussion, the family agrees to serve the evening meal in two choices. They all eat together and enjoy their meals equally. There is occasional grumbling about the extra work involved, so soon the whole family is eating vegetarian meals a few times a week to save on cooking. A new norm is established. Once the 25 per cent critical threshold is reached, a determined minority can change the opinion of a whole population.

Evidence shows that environmentalism – as we have practised it for 50 years, focused on protecting nature, building awareness, and assuming different levels of sacrifice (to save nature or the climate for us "bad humans") – hits a glass ceiling at some 15 per cent of citizens. These are the relatively well-educated, younger, generally middle-class, urban populations. They are willing to "save the planet" and "stop flying" or pay more for fuel, food, or housing to succeed. But the vast majority of citizens are indifferent to the planet. This majority wants to go on with their lives. As long as sustainability is equal to sacrifice, we believe that the exponential journey we need to embark on is dead in the water. Our only way to succeed is to prove that the journey back into a safe operating space on Earth is one that benefits you, me, our children, and their children. And not only economically, but also in terms of health and security.

This does not mean that environmentalism does not have a role to play. On the contrary, in order to "move mountains" you need the pioneers, the planetary sherpas who guide us on our journey, keep the debate vibrant and alive, and remind us all the time that there is a different world out there. So, we need XR, Greenpeace, and the razor-sharp voice of Greta, more than ever, as a continuous and strong drumbeat, reminding the world what is at stake.

The climate protests, media attention, planetary emergency declarations, and shocks of extreme weather events have thrown open the so-called Overton Window of political possibilities – the range of politically acceptable policies at any given time. It is no longer a question of whether we are moving towards a fossil fuel-free world; the question is whether we are doing it fast enough.

2. The political tipping point

In June 2019, the UK Parliament reached its lowest point in the country's bitter divorce battle with the European Union. The prime minister at the time, Theresa May, had failed to negotiate an agreement and had no choice but to resign. No one knew what would happen next. Deep distrust and anger permeated both of the largest political parties. Yet, somehow, amid the Brexit chaos, acrimony, and dysfunction, something truly remarkable happened. On 27 June 2019, Parliament agreed to amend the 2008 Climate Change Act to enshrine in law a target to reach net zero emissions by 2050. The United Kingdom became the first G7 nation to lock such a deep commitment into the statute books.

We were astounded that things could move so fast. We had thought that, given the Brexit distraction, any meaningful long-term policy was not possible – that the Overton Window had been bricked up. But the lead-up to this event was crucial. The climate strikes had exploded around the world, attracting political and media attention. XR was organizing "die-ins", in which protesters lay motionless in roads, shopping centres, and town squares. In October 2018, the Intergovernmental Panel on Climate Change had released a report that found that net zero by 2050 was not only essential but also feasible. And in May 2019, the United Kingdom's Committee on Climate Change had released its report "Net zero. The UK's contribution to stopping global warming". This argued that net zero was feasible for the United Kingdom, but it also went further. According to the report, the cost of the transition was likely to be the same as the estimated cost of reducing emissions by 80 per cent – 1 to 2 per cent of GDP – which the government had already committed to through existing legislation. The political conversation had shifted rapidly, from net zero emissions by 2050 being "impossible" to being "inevitable". In February 2020, the new prime minister, Boris Johnson, implored more major nations to adopt a 2050 target. The United Kingdom is playing a canny game here. A global economy without fossil fuels is inevitable, and the United Kingdom is adopting the same strategy it took 200 years ago during the industrial revolution: first-mover advantage.

More than 120 countries are now discussing a 2050 climate target. Denmark, France, New Zealand, and Sweden have also now adopted a net zero target in law. Sweden is aiming to reach the target by 2045. Some plan to reach net zero significantly earlier. In Norway and Uruguay, political leaders believe it is possible by 2030. Finland has its sights on 2035, while Iceland is aiming for 2040.

The European Union is committed to adopting a European Green Deal. Such a deal promises big investment in infrastructure and long-term planning to reach net zero by 2050. Just one country, Poland, is refusing to play ball. This is not likely to dent progress, though. The strong turnout for national Green parties in the European elections in 2019 has given the European Union the confidence and mandate to drive through this ambitious deal.

Perhaps most remarkably, in September 2020, China's president stunned everyone when he announced China would be net zero by 2060 at the latest. Two months later, Joe Biden won the US presidential election and vowed to launch a USD 2 trillion climate plan that will transform the foundation of the country's economy. This is big news. The global economy is driven by the United States, China, and Europe – let's call them the G3. We should not underestimate what a game-changing moment this is for our planet.

Some countries are going even further. They are rejecting outdated, socially destructive, Holocene economic dogma. In 2019, the prime minister of New Zealand, Jacinda Ardern, published the country's first "well-being budget" – a budget from the treasury that goes beyond a singular focus on GDP. Ardern has teamed up with Iceland's prime minister, Katrín Jakobsdóttir, and Scotland's first minister, Nicola Sturgeon, to put social and ecological well-being at the heart of their economic policies. These countries, along with Wales, have formed a new loose alliance: the Well-being Economy Governments, or WEGo. This is a breakthrough moment for ideas that can now be turned into political reality. Perhaps this group will grow in size and influence and eventually replace the G20.

It is not just smaller economies taking bold steps into the future. We have come to a political tipping point. Europe has put in place the steps needed to divorce itself from fossil fuels. This sends three unmistakable signals. To investors, it advises, "Get out of fossil fuels before the carbon bubble bursts." To companies operating in Europe that use fossil fuels, it says, "Switch to alternatives." And to innovators within and outside Europe, it encourages, "Do what you do best."

The absolute best political strategy is to find consensus across the political divide on a target and incrementally ratchet it up. Politicians have already begun to put in place policies that send strong signals to markets. The playing field is changing. These signals are set to cascade through the world's financial system, one of the most powerful forces for change on the planet.

3. The economic tipping point

The economic tipping point is more accurately described as a bag of tipping points. Growth in solar and wind power is probably the most visible one. Wind and solar are doubling every four or five years or so. If we keep up this pace, then at least half of global electricity will come from these sources by 2030. The numbers are astounding. Take photovoltaic cells used to produce solar energy, for example. Between 1975 and 2016, the price of photovoltaic cells plummeted 99.5 per cent. Every time the installed capacity doubled, the price fell by 20 per cent. For the past decade, photovoltaic installations grew 38 per cent every year. How can the fossil fuels possibly compete?

The sector will reach an unstoppable tipping point once renewable energy production is cheaper and more profitable than energy production from fossil fuels. In fact, this tipping point has already been reached. A global threshold is here. In September 2019, wind and solar power dropped below the price of coal in most of the world: in India, solar is now about half the price of coal. Wind and solar are already cheaper than gas in China, and could be cheaper than coal by 2026. At this rate, the Chinese and Indian governments would be crazy to build more power stations. The savings made by switching to wind and solar sooner rather than later speak for themselves. Of course, moving away from fossil fuels brings other benefits, too. One is breathable cities; the other is geopolitical. China imports coal, but manufactures and exports solar panels and wind. A rapid switch carries with it an additional economic kick and the promise of energy security for ever.

Financial analysts are still trying to get their heads around these new dynamics. Many economic scenarios show the future as an incremental continuation of the past – economists really struggle with tipping points. But we can see that something else is happening here. The current economic dynamics are reminiscent of the major phase changes that occurred at the start of the industrial revolution. No one wants to still be building canals when trains arrive. Perhaps the most telling sign is the fossil fuel sector's performance in the stock market index of largest companies by market capitalization, the S&P 500. In 2011, fossil fuels made up 12 per cent of the index. Now, they sit at just 4 per cent. According to *The Wall Street Journal*, energy – excluding renewables – was the S&P 500's worst-performing sector in 2014, 2015, 2018, and 2019. In April 2020, the pandemic caused the price of oil in the United States to crash below zero. Oil suppliers were forced to pay companies around USD 30 per barrel to take their oil.

It is clear that an increasing number of institutions are now divesting from fossil fuels. The value of divestments has already accelerated past USD 14 trillion. So, how far are we from another economic tipping point? Socially responsible investors have leverage. According to some research, only 10 to 20 per cent of investors can initiate a chain reaction that bursts the carbon bubble.[26, 27] More recent research by Ilona Otto and colleagues suggests the figure could be lower still. "A growing number of analysts believe a financial bubble is emerging that could burst when investors' belief

------

26 This neatly conforms to the 80/20 Pareto principle, a general heuristic that 80 per cent of an effect comes from 20 per cent of the causes. So, 20 per cent of customers may drive 80 per cent of sales, say. This idea is sometimes co-opted by those looking at tipping points in networks of people.
27 The name derives from an event called the South Sea Bubble. In the 1700s, greedy investors in England's South Sea Company anticipated huge profits from the slave trade with South America. This pushed share prices sky high, but the profits never materialized. The financial bubble eventually burst, bringing financial ruin to many institutions and individuals.

in carbon risk reaches a certain threshold. Simulations show that just 9 per cent of investors could tip the system, inducing other investors to follow."

4. The technological tipping point

The fourth and final tipping point is the technological revolution. An all-pervasive narrative informs us that technology, digital disruption, the fourth industrial revolution, whatever you want to call it, is barrelling towards us. While the past few decades have undoubtedly been disruptive, tech evangelists, media pundits, corporate consultancies, and government ministers tell us we ain't seen nothing yet.

Bill Gates, Mark Zuckerberg, and others are clearly in on the game. They are investing billions in innovation for planetary stewardship through a new initiative, Breakthrough Energy. But this is about investing in game-changing technology that may be a decade away from reaching consumers and scaling rapidly. We really need to focus on the near term. There are several technologies that are already scaling. These breakthroughs have reached an inflection point and will sweep through economies in the next decade. We have already covered renewable power. Electric vehicles are next. Electrical and plug-in hybrid passenger vehicle sales are growing exponentially. Today, they have an annual growth rate of 50 per cent.

The right economic policies will keep the world moving in the right direction at this breakneck speed. Norway has already shown the way. Thanks to strong policies, half of all new cars in the country are now electric or hybrid. If others adopt Norway's policies to grow electric vehicle sales 33 per cent every year, then by 2028 half of all cars will be electric, and the figure will be close to 100 per cent by the early 2030s. Many major economies and cities have announced incoming bans on new fossil-fuelled vehicles in the past few years. Big logistics companies such as Amazon and UPS, the United States postal giant, are committing to 100 per cent electric vehicles. In 2020, UPS ordered 10,000 electric vans and bought a chunk of UK electric vehicle pioneer Arrival. We cannot take any of these trajectories for granted, though; we still need strong support from lawmakers to lock them in.

In the next decade, digitalization will support a profound shift from ownership to usership. We are entering a sharing economy. Take car sharing, for example. Most vehicles spend 95 per cent of their time on parking lots. While autonomous vehicles are perhaps a decade away, car sharing is about to get a whole lot easier. Keys can now be shared digitally using mobile phones, thus creating one-click solutions to borrowing a car. These technologies will open up whole new business models.

E-commerce is transforming businesses and annihilating slow movers; online learning is transforming education; health is a new frontier, with promises of telemedicine and online diagnosis (remote precision surgery, where the surgeon is in a theatre in a different location from the patient, is already possible); and public administration is being overhauled through e-governance and e-payments. But we will see the biggest digital disruptions in the next decade in industry, agriculture, and finance. The 2020s will see the circular economy go mainstream, driven by digitalization.

Grocery shopping will change dramatically in the next decade, as giants such as Amazon and Alibaba stalk prey in this sector. Amazon has bought Whole Foods in the United States, and Alibaba is opening supermarkets across China. This could go hand in hand with a food revolution to nudge people to eat more healthy diets. A bunch of start-ups are already tackling food waste in restaurants, cafes, and supermarkets, using apps to connect hungry customers to cheap food. The potential is huge: if food waste were a country, it would have the third highest greenhouse gas emissions behind China and the United States. In farming, a combination of satellites, mobile phones, and drones are bringing precision agriculture and microweather forecasting to field level. And drones have been designed to plant trees more than 100 times faster than traditional methods.

We are already seeing major shifts in e-finance, too. Take Alibaba's Ant Financial, an online consumer bank with 800 million customers. Ant Financial has secured more than half of China's USD 13 trillion online payment market. More than 300 million people have signed up for Ant Forest, its online banking system, to nudge customers to choose low-carbon behaviour, such as using public transport, by giving them rewards and allowing them to compete for rewards with others in their social networks. The Swedish start-up Trine provides investors with a simple way to invest large or small sums in renewables in Africa. By combining a simple online investment platform with third-party verification provided by satellites, Trine overcomes a major hurdle for investment in Africa: perception of corruption.

Tech companies, both big and small, are eyeing the weakest, most vulnerable parts of the finance sector for opportunities to pounce. This momentum is unstoppable. We do still have some nagging doubts, though. Technology is a wild card. We have not been short of technological innovation in the past three decades, and emissions went up, not down. From nanotechnology and 3D printing to AI, automation, and algorithmic surveillance, the technology sector's disruption could work to restabilize the planet or further destabilize it. It could go either way.

Needless to say, with 60 per cent of the world now online, the technology sector is disproportionately influential. To ensure the industry is a force for good, world leaders such as Google, Amazon, and Apple need to commit to genuine planetary stewardship. While all have grand visions to conquer the world, few include restabilizing Earth as a priority. Ultimately, this means ditching the idea that technology is neutral, and instead actively using it to support societal goals, not undermine them.

In the 2020s, technology disruption is assured one way or the other. Within the industry, we can sense a profound shift. Apple, one of the world's most valuable companies, has committed to implementing a 100 per cent circular economy within its supply chain. Although they are not sure if they can actually get to 100 per cent, they know that they can get close. Companies who do not embrace the change feel the heat from their staff. Following Jeff Bezos's conspicuously low-level embrace of planetary stewardship, Amazon employees became increasingly vocal, until he eventually announced more sweeping measures within the company plus a USD 10 billion philanthropic fund. But Google, Amazon, Facebook, Apple, and others can and must do more. They need to look beyond philanthropy, beyond their own emissions, beyond the impact of their supply chains, and look directly at the consumers who use their products. Ultimately, we need a social contract between these tech platforms and their consumers to mutually support behavioural change to meet societal goals.

Social, political, economic, and technological tipping points are our superpowers. If just one of these four goes mainstream, then we substantially increase our chances of restabilizing Earth. Our best chance, though, is if they combine. This is already happening. Digital technology has fanned the flames of the school strikes and XR, building a global movement in months. This is influencing conversations everywhere. When we talk to politicians, business people, bureaucrats, everyone – without fail – references the new public mood. Every single one articulates the deep emotional impact from hearing our children's howl of pain. As lawmakers flex their muscles and move into uncharted territory of "net zero by 2050", the markets, Earth's proto-consciousness, start twitching, arching, and poising themselves for the moment to react. The pandemic recovery gives governments unexpected influence over the direction of the economy in the next decade. This influence must be channelled into the Earthshot mission.

WISE EARTH

There is something so strange and
beautiful about the presence of human minds
on Earth, about the fact that this one bit of
biosphere is awake and aware and curious
about everything, stuck in the gutter,
perhaps, but looking up at the stars.

DAVID GRINSPOON
EARTH IN HUMAN HANDS, 2016

We started this book hurtling down a mountain road at night, without headlights and without safety barriers or guardrails to demarcate where the side of the road ends and the precipitous drop begins. Scientific understanding is now allowing us to build guardrails for the first time. For several years, we have been working on three related research programmes: the World in 2050 project, Future Earth, and the Global Commons Alliance. In different ways, these projects and our own institutes are looking at the state of the planet now, and whether or not there are pathways towards a prosperous and just future for all within a safe operating space on Earth. The evidence at hand shows that there are. In fact, by 2030, we will know what our chances will be of eventually reaching this destination. By then, will we have cut global emissions of greenhouse gases by half? Halted the shocking loss of nature? Arrested deforestation? If we come to our senses now, the answer is "yes". And, we should know by as early as 2050 if the rate of change of Earth's life support system is stabilizing: that is, whether we are moving towards a safe landing zone, where we have harmonized our human enterprise with Earth's capacity to host us. If we are on target, by then, the ozone hole over Antarctica will have shrunk back close to a safe level. Greenhouse gas emissions should be tiny and balanced by our drawdown out of the atmosphere to deep underground. Ocean acidification may be stabilizing. Dead zones around the coasts could have recovered. Forests will be growing, not shrinking, and ecological collapse will have been averted. And by 2100? We may know by then if we are in this for the long haul.

We have come a long way.

In Act I, we told the story of the revolutions that made this planet: the emergence of life propelled our planet to a new stage, Earth 2.0 with a biosphere, then Earth 3.0 when photosynthetic life emerged, releasing oxygen to the atmosphere and creating the conditions for Earth 4.0 – the emergence of the first complex multicellular life some 540 million years ago. During the hundreds of millions of years that followed, an extremely warm hothouse Earth dominated, with mean average temperatures 5 to 10°C (9 to 18°F) warmer than our pre-industrial average of 14°C (57°F), little ice on the surface of our planet, and sea levels at least 70 metres (230 feet) higher than today, until the continents ripped apart and collided once more. This accelerated weathering, pulling carbon dioxide out of the air, until levels dropped to reach a threshold of approximately 350 parts per million (ppm). As carbon dioxide levels fell further, large ice sheets built up. Earth entered a curiously unstable ice age cycle, forcing rapid evolution of the brain of a remarkable mammal. This set the scene for the Holocene, the notably stable past 12,000 years, which began with one cultural revolution – agriculture – and ended with two – scientific and industrial.

In Act II, we told the story of arguably the three most important scientific insights of the past three decades. First, we are now in an entirely new geological epoch: the Anthropocene. Second, the Holocene is the only known state that can support our civilization. And third, we risk crossing dangerous tipping points if we wake the sleeping giants of the Amazon, Antarctica, Greenland, tundra, and ocean currents. Once one tipping point is crossed, it may cause a domino effect, pushing Earth back to the hothouse. The planet can cope with this state – it has been there before –

but it would crush our civilization and with it any aspirations of good lives for future generations on Earth.

These insights from academia demand a new world view as profound as those ushered in by the discoveries of Copernicus and Darwin: the fact that Earth orbits the Sun and the theory of evolution by means of natural selection, respectively. Both these insights challenged the institutions that had wielded unassailable power until then, shaking them to their core. The Anthropocene is destined to do the same. It is this generation's Copernican or Darwinian moment. It forces us to acknowledge that we live on a finite planet, pushed to its limits. We are approaching the edge of the lily pond.

In response to these scientific insights, we have now taken the first tentative steps to identify nine planetary boundaries to define a safe operating space for humanity. But Earth has already crashed through four of the nine boundaries, relating to climate change, collapse of biodiversity, deforestation, and our over-use of nutrients in farming. This is now a planetary emergency.

Finally, in Act III, we described the Earthshot mission and argued that every child's birthright is a stable, resilient planet. This is why our children are reading scientific articles and walking out of school to demand action. We all must become planetary stewards and protect our global commons. Central to this are finding a common human identity and opening up our eyes to the new world we have created. We face catastrophic risks, but luckily the solutions exist and offer prosperity, security, and equity.

Throughout this book, we have frequently used the term "we". Sometimes, we simply mean "us", as authors. Other times, we mean our species. Our academic colleagues are quick to flare up and say there is no "we". Subsistence farmers in sub-Saharan Africa did not cause the Great Acceleration that pushed us into the Anthropocene. It was Western elites and capitalism, they say. This is true, but we (authors) say that acknowledging a common human identity beyond the city we grew up in and beyond a nation state boundary is an important step on the journey to planetary stewardship: we are all on this unstable planet together. In some ways, planetary stewardship is already here, in the schoolchildren marching, in the CEOs of multinational corporations demanding tougher action, in the hearts of the women who run the European

Central Bank and the International Monetary Fund. It is just unevenly distributed. In the coming years, we hope we will find a collective "we" emerging.

In order to fully achieve planetary stewardship, the world must change course. Our Earthshot mission is to restabilize Earth, but we have little time left to act. However, if we respond now, the best science says we can succeed. If the bath is overflowing, do we argue about who turned on the taps? Do we build an elaborate piping system to funnel away the excess water? Or do we turn off the tap and take out the plug? The answer is simple. In the face of big risk, we choose the safest route. Now is the time to end our dependence on fossil fuels and embark on a sustainable Earth 5.0.

First, we need to set a few things straight. Our economic system is based on the assumption of infinity: that there are infinite resources, such as the atmosphere and the ocean, with infinite reservoirs for waste products, and that there is infinite capacity for growth. This is obviously wrong. And it is also potentially the heart of the reason why transitioning away from the old unsustainable economic paradigm is so painful for many. In the beginning, some 240 years ago, Adam Smith set the stage for capitalism. He was followed by a plethora of scholars and thinkers, from John Maynard Keynes to Milton Friedman. Collectively, they provided the intellectual support for the market economy and globalization as we know it today. This was based on a world view that the economic world was truly small and could generate wealth, on what appeared to be a big planet with infinite resources of soils, forests, metals, atmosphere, water and nutrients. It is this free subsidy – providing biomass, natural resources, and waste disposal for free – that is the basis for our USD 86 trillion global economy.

But the global economy has now unequivocally reached a saturation point for the planet. We have been in this state for at least 30 years, with no correction in the course of the global economy. This is the drama. Our current economic paradigm hit a hard wall and became obsolete several decades ago. The proof is in the transgression of planetary boundaries: in the cracks appearing in the Antarctic ice sheet; in the Amazon's failing ability to store carbon; in the melting permafrost in the far north; in the fires, droughts, locust plagues, and floods; in the coral reefs our children may not see; and in the exponential spread of a pandemic through densely populated cities.

What we need is a new paradigm where we no longer accept the pursuit of economic growth at the expense of the planet. We need to reconnect people with the planet, and seek prosperity and equity within a safe operating space on Earth. Luckily, the tools we need are largely available. We are talking about regenerative agriculture, nature-based solutions, circular economic models, science-based targets for business, and collective governance of our global commons at all scales, from local wetlands to the ice sheets and the ocean. These are all tested solutions that are available as islands of success. They exist in an ocean of indifference, or at worst, face resistance or ignorance, but they are there.

This suggests that the economy can develop within the safe operating space marked by the nine planetary boundaries. There, the economic system – defined as generating human wealth, harmony, and security – can infinitely regenerate, for all intents and purposes, like the biosphere. But only if we harness energy from the Sun and close the loop to recirculate steel, cement, plastic, aluminium, and other materials.

Some claim that this strategy is too simple. That the second law of thermodynamics always prevails: creating an ordered civilization – order from chaos – necessarily means creating more waste. This is true. There will always be a marginal footprint from our human enterprise. At the same time, the Earth system has proven how resilient it is. It has shown a remarkable ability to dampen and absorb human abuse so far during our Anthropocene journey. If we are able to safeguard the remaining resilience of Earth, particularly through a zero loss of nature law, then there is a good chance that Earth can cope with a continued residual environmental impact from our world, increasing the chances of reaching Earth 5.0. Consequently, the planetary boundaries are set at the level – 350 ppm of carbon dioxide, for example – of a precautionary assessment of what Earth can cope with without drifting onto a hothouse trajectory. Furthermore, there is a great deal of evidence that we already have solutions to regenerate life support systems. We can be a positive force for the biosphere. We can be "nature positive" through sustainable agriculture, restoring ecosystems, cleaning up lakes. Our capacity for creativity, innovation, and knowledge generation may be infinite or it may not. But we are certainly not going to exhaust all possibilities any time soon. All our efforts need to go into building and distributing wealth by decarbonizing energy production and closing the loops as soon as possible.

Our current narrative comes to an end at a moment of high drama. Unlike a Hollywood blockbuster, which draws to a neat conclusion with a nice resolution and a happy/relieved hero/heroine who has learned something new, this story is more like a Netflix binge watch, ending on a cliffhanger.

We have 10 years. A decade.

We are very aware of the risk of deadlines such as "Only 10 years to save the planet". Some say that it is never too late, or that the changes will be gradual and we can adapt. But this does not reflect the state of knowledge. That said, let us be crystal clear. We are, of course, not suggesting that Earth will fall off a cliff on 1 January 2031. But science is telling us that if we are not able to turn things around over the coming decade – cutting global emissions by half and halting the loss of nature – then we face a great risk of pressing the "on" buttons of irreversible change on Earth. We risk crossing unstoppable tipping points that will lock us onto a path towards a hothouse Earth state.

If the West Antarctic ice sheet disintegrates or the Amazon starts belching carbon, both of which are possible within the next two decades – the former is very likely crossing a tipping point in this respect as we speak – human societies will probably be able to adapt. Huge investment in sea walls to protect vulnerable cities will need to be accelerated, and carbon capture and storage solutions will need to scale rapidly, while reducing greenhouse gases at an even faster rate than the breakneck 7 per cent per year. Even if we manage to achieve this, societies will have to adapt continually for centuries. And climate change is likely to accelerate. Once we take into consideration the usual economic and political turbulence, and the occasional pandemic, you can see we are heading towards the rapids. The global markets hate uncertainty, yet our future will be increasingly driven by it. Anyone who believes it will be possible to restabilize Earth, and bring people out of poverty, while the markets panic is living in a fantasy world. Most people on Earth were affected in some way by the economic fall-out of COVID-19. These impacts will last decades, if not generations. This is small fry compared with what is coming.

The Earthshot mission, therefore, is not a mission towards a utopian dream. It is about economic stability within a stable, thriving biosphere. This is not something we are aiming for; it is something we are navigating back to. It is not, therefore, an environmental agenda. It is not about protecting nature, but about our own prosperity and our ability to navigate a non-linear world, as well as, potentially, the ability of democratic systems to prevail. Because the more we delay the transformation, the more we risk needing to implement unpopular draconian measures. If we want some degree of freedom left, and the room for citizen influence and peaceful collective action, then we need to act now. If we lose this window of opportunity, then either we will fail (and face the consequences) or we will enter a phase of overbearing top-down rule in order to curb emissions.

Fortunately, we have a number of success stories to fall back on when we feel frustrated, angry, and ready to give up. We have already begun to repair the hole in the ozone layer. We have dealt with acid rain. We have banned nuclear testing in the atmosphere. We, as a civilization, have survived global pandemics.

Keeping on course ultimately means political leadership and cooperation. A strength, perhaps the greatest strength, of democratic societies is to find a way for people with opposing views to agree a path forward. And undoubtedly our greatest asset as a species is our ability to cooperate. This gives us our superpower. We need to find it again. Planetary stewardship is far more likely to emerge if there is trust in political institutions to make long-term decisions.

We can all use our superpowers to make this happen. We are all agents of change. Never have individuals had the power to influence so much. And it takes only a small handful of committed individuals to tip the balance in a community or society.

You have consumer power. Voting power. Network power.

Use it.

If the world can find the means to cooperate at the planetary scale, it will be humanity's greatest success: a restabilized planet. Perhaps we might finally deserve the *sapiens* or "wise" element in our Latin name.

This achievement may be even more profound than we realize. David Grinspoon, an astrobiologist and former NASA researcher, looks at the Anthropocene from a cosmological perspective and wonders, are we missing something big?

In the past three decades, while Earth system researchers came to a new understanding of our planet, astronomers were busy identifying more than 4,000 exoplanets. This has been a remarkable period in astronomy. We can now identify features about these exoplanets. Are they rocks or gas giants? Do they have some sort of atmosphere or not? In the next decade, as the Earthshot mission gathers pace, astronomers may be able to detect signs of liquid water and the composition of the atmosphere: does it contain oxygen, nitrogen, methane, water, carbon dioxide?

From this information we could work out each planet's current evolutionary phase, just like our own planet: is it lifeless, like the Hadean aeon? Does it have a life support system of some description? Are there telltale signs of a biosphere? Perhaps it has reached a similar state to Earth 2.0, with life but no photosynthesis, where an early, simple biosphere interacts with the planet's physical systems. Is the atmosphere rich in oxygen, indicating single-celled photosynthesis at work, like Earth 3.0, and a biosphere acting more strongly on physical and chemical processes of the planet? Is complex life possible, like Earth 4.0, where life unconsciously influences the habitability of the planet?

But what if there is a fifth phase in planetary evolution? A phase where the biosphere consciously manages the planet for habitability. This is a logical leap based on what has happened in recent decades. Through science, a planetary awareness has emerged. Through the global economy and technological innovation, a mechanism for influencing the habitability of Earth has also emerged. It is not too much to believe that we might, one day, consciously manage the planet's physical, chemical, and biological systems to maintain habitability – to maintain a thriving, rich, diverse biosphere in some sort of harmony. Grinspoon proposes that this phase in Earth's evolution, Earth 5.0, marks the arrival of some sort of planetary intelligence. He says it could be termed *Terra sapiens*: wise Earth.

Let's not get ahead of ourselves. The rate of change of Earth's life support system is accelerating. A distant civilization inhabiting a star system far, far away would be able to detect signs of the Great Acceleration over the past 70 years. The signal would be unmistakable, and it would probably be seen as a rupture. Viewers might conclude they are witnessing some sort of cataclysmic disruption on our planet. But what? A volcanic eruption from deep within the core? A solar flare from the planet's nearest star? An asteroid collision? A gamma ray burst from the collision of two black holes or neutron stars? An evolutionary leap? Or perhaps a technologically advanced civilization stretching beyond its reach ... ?

Like a doctor looking at a patient's vital signs on a monitor, alien observers might reasonably wonder, what happens next? Will the exponential curves keep rising and become even more disruptive and chaotic as the whole system amplifies the disturbance? Or will the rate of change of the planet's life support system stop fluctuating wildly and settle into a more stable state? This might be interpreted as a flickering of consciousness, of intent, at a planetary level.

We are a long way from this lofty goal. But here's the thing: our research shows that stability is achievable within 30 years. A nature-positive pathway is not wishful thinking. In three decades, Earth's natural ecosystems can be stronger, more resilient, and, critically, more extensive than today. Imagine that.

Our economies, too, can be strong and more resilient, more able to roll with the punches and rebound from shocks. Imagine that.

It is time to rebuild Earth's resilience with an economy that not only operates within planetary boundaries but also actively enhances the resilience of Earth. This is the Earthshot mission. This is our mission for the next decade. We need everyone on board.

Let's leave our children nothing. No greenhouse gas emissions. No biodiversity loss. No poverty. Let's leave them what we inherited: a stable, resilient planet. Not to save our planet. But to save ourselves and our future on Earth.

SOURCES

ACT I

Chapter 1
J. D. Archibald, *Dinosaur Extinction and the End of an Era: What the Fossils Say*, Columbia University Press, 1996.

A. D. Barnosky et al, "Has the Earth's sixth mass extinction already arrived?", *Nature*, 471 (7336), 2011.

Y. M. Bar-On, R. Phillips, and R. Milo, "The biomass distribution on Earth", *Proceedings of the National Academy of Sciences*, 115 (25), 2018.

S. Boon, "21st century science overload", *Canadian Science Publishing*. Available at: blog.cdnsciencepub.com/21st-century-science-overload/

T. W. Crowther et al, "Mapping tree density at a global scale", *Nature*, 525 (7568), 2015.

M. S. Dodd et al, "Evidence for early life in Earth's oldest hydrothermal vent precipitates", *Nature*, 543 (7643), 2017.

J. G. Dyke and I. S. Weaver, "The emergence of environmental homeostasis in complex ecosystems", *PLoS Computational Biology*, 9 (5), 2013.

G. Feulner, "The faint young Sun problem", *Reviews of Geophysics*, 50 (2), 2012.

P. F. Hoffman et al, "A Neoproterozoic Snowball Earth", *Science*, 281 (5381), 1998.

A. E. Jinha, "Article 50 million: an estimate of the number of scholarly articles in existence", *Learned Publishing*, 23 (3), 2010.

R. Johnson, A. Watkinson, and M. Mabe, "The STM report: an overview of science and scholarly publishing", 2018.

J. L. Kirschvink, "Late Proterozoic low-latitude global glaciation: the Snowball Earth", *The Proterozoic Biosphere: A Multidisciplinary Study*, J. W. Schopf and C. Klein (Eds), Cambridge University Press, 1992.

M. LaFrance, M. A. Hecht, and E. L. Paluck, "The contingent smile: A meta-analysis of sex differences in smiling", *Psychological Bulletin*, 129 (2), 2003.

T. Lenton, *Earth System Science: A Very Short Introduction*, Oxford University Press, 2016.

J. E. Lovelock and L. Margulis, "Atmospheric homeostasis by and for the biosphere: the Gaia hypothesis", *Tellus*, 26 (1–2), 1974.

T. W. Lyons, C. T. Reinhard, and N. J. Planavsky, "The rise of oxygen in Earth's early ocean and atmosphere", *Nature*, 506 (7488), 2014.

C. R. Marshall, "Explaining the Cambrian 'explosion' of animals", *Annual Review of Earth and Planetary Sciences*, 34 (1), 2006.

M. Maslin, *The Cradle of Humanity: How the Changing Landscape of Africa Made us so Smart*, Oxford University Press, 2017.

C. Patterson, "Age of meteorites and the Earth", *Geochimica et Cosmochimica Acta*, 10 (4), 1956.

M. R. Rampino and S. Self, "Volcanic winter and accelerated glaciation following the Toba super-eruption", *Nature*, 359 (6390), 1992.

R. M. Soo et al, "On the origins of oxygenic photosynthesis and aerobic respiration in Cyanobacteria", *Science*, 355 (6332), 2017.

J. Tyndall, *Contributions to Molecular Physics in the Domain of Radiant Heat: A Series of Memoirs Published ...* Longmans, Green, and Company, 1872.

Chapter 2
S. Barker et al, "800,000 years of abrupt climate variability", *Science*, 334 (6054), 2011.

J. Croll, "XIII. On the physical cause of the change of climate during geological epochs", *The London, Edinburgh, and Dublin Philosophical Magazine and Journal of Science*, 28 (187), 1864.

W. Köppen and A. Wegener, *The Climates of the Geological Past*, Borntraeger, 1924.

M. Milankovic, *Canon of Insolation and the Ice-Age Problem*, Agency for Textbooks, 1998.

J. R. Petit et al, "Climate and atmospheric history of the past 420,000 years from the Vostok ice core, Antarctica", *Nature*, 399 (6735), 1999.

M. Willeit et al, "Mid-Pleistocene transition in glacial cycles explained by declining CO2 and regolith removal", *Science Advances*, 5 (4), 2019.

Chapter 3
A. Bardon, "Humans are hardwired to dismiss facts that don't fit their worldview", *The Conversation*. Available at: theconversation.com/humans-are-hardwired-to-dismiss-facts-that-dont-fit-their-worldview-127168

B. de Boer, "Evolution of speech and evolution of language", *Psychonomic Bulletin & Review*, 24 (1), 2017.

P. B. deMenocal, "Climate and human evolution", *Science*, 331 (6017), 2011.

M. González-Forero and A. Gardner, "Inference of ecological and social drivers of human brain-size evolution", *Nature*, 557 (7706), 2018.

B. Hare, "Survival of the friendliest: *Homo sapiens* evolved via selection for prosociality", *Annual Review of Psychology*, 68 (1), 2017.

B. Hare, V. Wobber, and R. Wrangham, "The self-domestication hypothesis: evolution of bonobo psychology is due to selection against aggression", *Animal Behaviour*, 83 (3), 2012.

F. Jabr, "Does thinking really hard burn more calories?", *Scientific American*. Available at: www.scientificamerican.com/article/thinking-hard-calories/

I. Martínez et al, "Communicative capacities in Middle Pleistocene humans from the Sierra de Atapuerca in Spain", *Quaternary International*, 295, 2013.

I. McDougall, F. H. Brown, and J. G. Fleagle, "Stratigraphic placement and age of modern humans from Kibish, Ethiopia", *Nature*, 433 (7027), 2005.

H. Mercier and D. Sperber, "Why do humans reason? Arguments for an argumentative theory", *Behavioral and Brain Sciences*, 34 (2), 2011.

A. Navarrete, C. P. van Schaik, and K. Isler, "Energetics and the evolution of human brain size", *Nature*, 480 (7375), 2011.

S. Neubauer, J.-J. Hublin, and P. Gunz, "The evolution of modern human brain shape", *Science Advances*, 4 (1), 2018.

I. S. Penton-Voak and J. Y. Chen, "High salivary testosterone is linked to masculine male facial appearance in humans", *Evolution and Human Behavior*, 25 (4), 2004.

T. Rito et al, "A dispersal of *Homo sapiens* from southern to eastern Africa immediately preceded the out-of-Africa migration", *Scientific Reports*, 9 (1), 2019.

M. R. Sánchez–Villagra and C. P. van Schaik, "Evaluating the self-domestication hypothesis of human evolution", *Evolutionary Anthropology: Issues, News, and Reviews*, 28 (3), 2019.

E. M. L. Scerri et al, "Did our species evolve in subdivided populations across Africa, and why does it matter?", *Trends in Ecology and Evolution*, 33 (8), 2018.

S. Shultz, E. Nelson, and R. I. M. Dunbar, "Hominin cognitive evolution: identifying patterns and processes in the fossil and archaeological record", *Philosophical Transactions of the Royal Society B: Biological Sciences*, 367 (1599), 2012.

S. W. Simpson et al, "A female *Homo erectus* pelvis from Gona, Ethiopia", *Science*, 322 (5904), 2008.

E. A. Smith, "Communication and collective action: language and the evolution of human cooperation", *Evolution and Human Behavior*, 31 (4), 2010.

E. I. Smith et al, "Humans thrived in South Africa through the Toba eruption about 74,000 years ago", *Nature*, 555 (7697), 2018.

C. Stringer, "The origin and evolution of *Homo sapiens*", *Philosophical Transactions of the Royal Society B: Biological Sciences*, 371 (1698), 2016.

G. West, *Scale: The Universal Laws of Growth, Innovation, Sustainability, and the Pace of Life in Organisms, Cities, Economies, and Companies*, Penguin Press, 2017.

M. Williams, "The ~73 ka Toba super-eruption and its impact: history of a debate", *Quaternary International*, 258, 2012.

B. Wood and E. K. Boyle, "Hominin taxic diversity: fact or fantasy?", *American Journal of Physical Anthropology*, 159 (S61), 2016.

R. W. Wrangham et al, "The raw and the stolen: Cooking and the ecology of human origins", *Current Anthropology*, 40 (5), 1999.

Chapter 4

J. Diamond, *The Third Chimpanzee: The Evolution and Future of the Human Animal*, Harper Perennial, 2006.

J. Diamond, "The worst mistake in the history of the human race", *Discover Magazine*, 1987.

J. W. Erisman et al, "How a century of ammonia synthesis changed the world", *Nature Geoscience*, 1 (10), 2008.

N. Ferguson, *The Square and the Tower: Networks and Power, from the Freemasons to Facebook*, Penguin Press, 2018.

J. Feynman and A. Ruzmaikin, "Climate stability and the development of agricultural societies", *Climate Change*, 84 (3), 2007.

A. Ganopolski, R. Winkelmann, and H. J. Schellnhuber, "Critical insolation–CO2 relation for diagnosing past and future glacial inception", *Nature*, 529 (7585), 2016.

Intergovernmental Panel on Climate Change, "Summary for policymakers", *Special Report on the Impacts of Global Warming of 1.5°C*, Intergovernmental Panel on Climate Change, 2018.

P. H. Kavanagh et al, "Hindcasting global population densities reveals forces enabling the origin of agriculture", *Nature Human Behaviour*, 2 (7), 2018.

S. A. Marcott et al, "A reconstruction of regional and global temperature for the past 11,300 years", *Science*, 339 (6124), 2013.

D. J. Markwell, *John Maynard Keynes and International Relations: Economic Paths to War and Peace*, Oxford University Press, 2006.

L. Phillips and M. Rozworski, *People's Republic of Walmart: How the World's Biggest Corporations Are Laying the Foundation for Socialism*, Verso Books, 2019.

V. Smil, *Growth*, The MIT Press, 2019.

W. Steffen et al, "Planetary boundaries: guiding human development on a changing planet", *Science*, 347 (6223), 2015.

W. Steffen et al, "The trajectory of the Anthropocene: The Great Acceleration", *Anthropocene Review*, 2 (1), 2015.

United Nations Department of Economic and Social Affairs, "Post-war reconstruction and development in the Golden Age of Capitalism", *World Economic and Social Survey 2017*, United Nations Department of Economic and Social Affairs, 2017.

C. N. Waters et al, "The Anthropocene is functionally and stratigraphically distinct from the Holocene", *Science*, 351 (6269), 2016.

R. Wilkinson and K. Pickett, *The Spirit Level: Why Equality Is Better for Everyone*, Penguin, 2010.

ACT II

Chapter 5

T. Lenton et al, "Climate tipping points – too risky to bet against", *Nature*, 575 (7784), 2019.

T. Lenton, "Early warning of climate tipping points", *Nature Climate Change*, 1 (4), 2011.

W. Steffen et al, "The trajectory of the Anthropocene: The Great Acceleration", *Anthropocene Review*, 2 (1), 2015.

Chapter 6

C. Folke et al, "Resilience thinking: integrating resilience, adaptability and transformability", *Ecology and Society*, 15 (4), 2010.

T. Fuller (Ed), *Gnomologia, Adagies and Proverbs, Wise Sentences and Witty Sayings, Ancient and Modern, Foreign and British*, Kessinger Publishing, 2003.

M. Gladwell, *The Tipping Point: How Little Things Can Make a Big Difference*, Back Bay Books, 2002.

T. Lenton et al, "Tipping elements in the Earth's climate system", *Proceedings of the National Academy of Sciences*, 105 (6), 2008.

J. E. Lovelock and L. Margulis, "Atmospheric homeostasis by and for the biosphere: the Gaia hypothesis", *Tellus*, 26 (1–2), 1974.

M. E. Mann, R. S. Bradley, and M. K. Hughes, "Northern hemisphere temperatures during the past millennium: inferences, uncertainties, and limitations", *Geophysical Research Letters*, 26 (6), 1999.

J. C. Rocha et al, "Cascading regime shifts within and across scales", *Science*, 362 (6421), 2018.

Chapter 7

R. J. W. Brienen et al, "Long-term decline of the Amazon carbon sink", *Nature*, 519 (7543), 2015.

D. W. Fahey et al, "The 2018 UNEP/WMO assessment of ozone depletion: an update", abstract #A31A-01 presented at the AGU Fall Meeting, 2018.

M. Grooten and R. Almond, *Living Planet Report 2018: Aiming Higher*, WWF, Gland, Switzerland, 2018.

B. Hönisch et al, "The geological record of ocean acidification", *Science*, 335 (6072), 2012.

Intergovernmental Panel on Climate Change, "Summary for Policymakers", *Climate Change 2013: The Physical Science Basis. Contribution of Working Group I to the Fifth Assessment Report of the Intergovernmental Panel on Climate Change*, Intergovernmental Panel on Climate Change, 2013.

Intergovernmental Science-Policy Platform on Biodiversity and Ecosystem Services, "Global assessment report on biodiversity and ecosystem services", IPBES, 2019.

T. E. Lovejoy and C. Nobre, "Amazon tipping point", *Science Advances*, 4 (2), 2018.

V. Masson-Delmotte et al, "Information from paleoclimate archives", in *Climate Change 2013: The Physical Science Basis. Contribution of Working Group I to the Fifth Assessment Report of the Intergovernmental Panel on Climate Change*, Cambridge University Press, 2013.

G. Readfearn, "Climate crisis may have pushed world's tropical coral reefs to tipping point of 'near-annual' bleaching", *The Guardian*, 2020.

J. Rockström et al, "A safe operating space for humanity", *Nature*, 461 (7263), 2009.

J. Rockström et al, "Planetary boundaries: exploring the safe operating space for humanity", *Ecology and Society*, 14 (2), 2009.

S. Solomon, "The mystery of the Antarctic ozone 'hole'", *Reviews of Geophysics*, 26 (1), 1988.

W. Steffen et al, "Planetary boundaries: guiding human development on a changing planet", *Science*, 347 (6223), 2015.

W. Steffen et al, "Trajectories of the Earth system in the Anthropocene", *Proceedings of the National Academy of Sciences*, 115 (33), 2018.

E. O. Wilson, *Half-Earth: Our Planet's Fight for Life*, W. W. Norton & Company, 2016.

World Meteorological Organization (WMO), "WMO provisional statement on the state of the global climate in 2019", *WMO Statement on the State of the Global Climate*, WMO, 2019.

Chapter 8

Club of Rome and the Potsdam Institute for Climate Impact Research, "The planetary emergency plan", 2019. Available at: clubofrome.org/publication/the-planetary-emergency-plan/

D. Coady et al, "Global fossil fuel subsidies remain large: an update based on country-level estimates", Working Paper no. 19/89, IMF, 2019.

W. Hubau et al, "Asynchronous carbon sink saturation in African and Amazonian tropical forests", *Nature*, 579 (7797), 2020.

Intergovernmental Panel on Climate Change, "Summary for policymakers", *Special Report on the Ocean and Cryosphere in a Changing Climate*, IPCC, 2019.

P. Milillo et al, "Heterogeneous retreat and ice melt of Thwaites Glacier, West Antarctica", *Science Advances*, 5 (1), 2019.

T. A. Scambos et al, "How much, how fast?: A science review and outlook for research on the instability of Antarctica's Thwaites Glacier in the 21st century", *Global and Planetary Change*, 153, 2017.

T. Schoolmeester et al, *Global Linkages: A Graphic Look at the Changing Arctic (rev. 1)*, UN Environment and GRID-Arendal, 2019.

A. Shepherd et al, "Mass balance of the Greenland Ice Sheet from 1992 to 2018", *Nature*, 579, (7798), 2020.

ACT III

Chapter 9

M. Carney, F. V. de Galhau, and F. Elderson, "The financial sector must be at the heart of tackling climate change", *The Guardian*, 2019.

E. Daly, "The Ecuadorian exemplar: the first ever vindications of constitutional rights of nature", *Review of European Community & International Environmental Law*, 21 (1), 2012.

L. Fink, "CEO letter", *BlackRock*. Available at: www.blackrock.com/uk/individual/larry-fink-ceo-letter

B. Gates, N. Myhrvold, and P. Rinearson, *The Road Ahead: Completely Revised and Up-to-Date*, Penguin Books, 1996.

G. R. Harmsworth and S. Awatere, "Indigenous Māori knowledge and perspectives of ecosystems", *Ecosystem Services in New Zealand–Conditions and Trends*, 2013.

D. Meadows, "Leverage points: places to intervene in a system", *Academy for Systems Change*. Available at: donellameadows.org/archives/leverage-points-places-to-intervene-in-a-system/

D. H. Meadows et al, *The Limits to Growth*, A report to the Club of Rome, 1972.

E. Ostrom et al, "Revisiting the commons: local lessons, global challenges", *Science*, 284 (5412), 1999.

E. Röös, M. Patel, and J. Spångberg, "Producing oat drink or cow's milk on a Swedish farm: environmental impacts considering the service of grazing, the opportunity cost of land and the demand for beef and protein", *Agricultural Systems*, 142, 2016.

S. Rotarangi and D. Russell, "Social–ecological resilience thinking: can indigenous culture guide environmental management?, *Journal of the Royal Society of New Zealand*, 39 (4), 2009.

G. Turner, "Is global collapse imminent? An updated comparison of *The Limits to Growth* with historical data", *MSSI Research Paper*, 4, 2014.

D. Wallace-Wells, *The Uninhabitable Earth: Life After Warming*, Tim Duggan Books, 2019.

E. O. Wilson, *Half-Earth: Our Planet's Fight for Life*, W. W. Norton & Company, 2016.

Chapter 10

K. Anderson, "Talks in the city of light generate more heat", *Nature News*, 528 (7583), 2015.

P. Hawken, *Drawdown: The Most Comprehensive Plan Ever Proposed to Reverse Global Warming*, Penguin, 2018.

C. Le Quéré et al, "Drivers of declining CO_2 emissions in 18 developed economies", *Nature Climate Change*, 9 (3), 2019.

E. Morena, *The Price of Climate Action: Philanthropic Foundations in the International Climate Debate*, Palgrave Macmillan, 2016.

J. Rockström et al, "A roadmap for rapid decarbonization", *Science*, 355 (6331), 2017.

Chapter 11

B. M. Campbell et al, "Agriculture production as a major driver of the Earth system exceeding planetary boundaries", *Ecology and Society*, 22 (4), 2017.

T. Lucas and R. Horton, "The 21st-century great food transformation", *The Lancet*, 393 (10170), 2019.

W. J. McCarthy and Z. Li, "Healthy diets and sustainable food systems", *The Lancet*, 394 (10194), 2019.

L. Olsson et al, "Land degradation", *Climate Change and Land*, Intergovernmental Panel on Climate Change, 2019.

J. Rockström and M. Falkenmark, "Agriculture: increase water harvesting in Africa", *Nature*, 519 (7543), 2015.

R. Scholes et al (eds), "Summary for policymakers of the assessment report on land degradation and restoration of the Intergovernmental Science-Policy Platform on Biodiversity and Ecosystem Services", IPBES, 2018.

M. Shekar and B. Popkin, *Obesity: Health and Economic Consequences of an Impending Global Challenge*, World Bank Publications, 2020.

W. Willett et al, "Food in the Anthropocene: the EAT–*Lancet* Commission on healthy diets from sustainable food systems", *The Lancet*, 393 (10170), 2019.

Chapter 12

M. Burke, S. M. Hsiang, and E. Miguel, "Global non-linear effect of temperature on economic production", *Nature*, 527 (7577), 2015.

A. Chrisafis, "Macron responds to gilets jaunes protests with €5bn tax cuts", *The Guardian*, 2019.

N. S. Diffenbaugh and M. Burke, "Global warming has increased global economic inequality", *Proceedings of the National Academy of Sciences*, 116 (20), 2019.

D. Hardoon, R. Fuentes-Nieva, and S. Ayele, "An economy for the 1%: how privilege and power in the economy drive extreme inequality and how this can be stopped", Oxfam International, 2016.

Intergovernmental Panel on Climate Change, "Summary for policymakers", *Climate Change and Land*, IPCC, 2019.

P. R. La Monica, "Warren Buffett has $130 billion in cash. He's looking for a deal", *CNN Business*, 2020.

I. M. Otto et al, "Shift the focus from the super-poor to the super-rich", *Nature Climate Change*, 9 (2), 2019.

Oxfam International, "Just 8 men own same wealth as half the world", 2018. Available at: www.oxfam.org/en/press-releases/just-8-men-own-same-wealth-half-world

T. Piketty, *Capital in the Twenty-First Century*, Harvard University Press, 2017.

A. Shorrocks et al, "Global wealth report 2019", Credit Suisse Research Institute, 2019. Available at: www.credit-suisse.com/media/assets/corporate/docs/about-us/research/publications/global-wealth-report-2019-en.pdf

World Food Programme, "Southern Africa in throes of climate emergency with 45 million people facing hunger across the region". Available at: www.wfp.org/news/southern-africa-throes-climate-emergency-45-million-people-facing-hunger-across-region

Chapter 13

F. Akthar and E. Dixon, "At least 36 people dead in one of India's longest heatwaves", *CNN*, 2019.

M. Artmann, L. Inostroza, and P. Fan, "Urban sprawl, compact urban development and green cities. How much do we know, how much do we agree?", *Ecological Indicators*, 96, 2019.

J. Drevikovsky and S. Rawsthorne, "'Hottest place on the planet': Penrith in Sydney's west approaches 50 degrees", *The Sydney Morning Herald*, 2020.

B. Eckhouse, "The U.S. has a fleet of 300 electric buses. China has 421,000", *Bloomberg*, 2019.

J. Falk and O. Gaffney et al, "Exponential climate action roadmap", Future Earth, 2018. Available at: exponentialroadmap.org/wp-content/uploads/2018/09/Exponential-Climate-Action-Roadmap-September-2018.pdf

T. Frank, "After a $14-billion upgrade, New Orleans' levees are sinking", *Scientific American*, 2019. Available at: www.scientificamerican.com/article/after-a-14-billion-upgrade-new-orleans-levees-are-sinking/

D. Hoornweg et al, "An urban approach to planetary boundaries", *Ambio*, 45 (5), 2016.

International Energy Agency, "Cities are at the frontline of the energy transition", 2016. Available at: www.iea.org/news/cities-are-at-the-frontline-of-the-energy-transition

S. A. Kulp and B. H. Strauss, "New elevation data triple estimates of global vulnerability to sea-level rise and coastal flooding", *Nature Communications*, 10, 2019.

J. Lelieveld et al, "Loss of life expectancy from air pollution compared to other risk factors: a worldwide perspective", *Cardiovascular Research*, 2020.

B. Mason, "The ACT is now running on 100 renewable electricity", *SBS News*. Available at: www.sbs.com.au/news/the-act-is-now-running-on-100-renewable-electricity

W. Rees and M. Wackernagel, "Urban ecological footprints: why cities cannot be sustainable – and why they are a key to sustainability", *Environmental Impact Assessment Review*, 16 (4–6), 1996.

H. Ritchie and M. Roser, "Urbanization", *Our World in Data*, 2018. Available at: ourworldindata.org/urbanization

D. Robertson, "Inside Copenhagen's race to be the first carbon-neutral city", *The Guardian*, 2019.

"Scientists disappointed with New Urban Agenda agreed on by nations at Habitat III summit", *International Science Council*, 2016. Available at: council.science/current/press/scientists-disappointed-with-new-urban-agenda-agreed-on-by-nations-at-habitat-iii-summit/

M. Sheetz, "Technology killing off corporate America: Average life span of companies under 20 years", *CNBC*, 2017. Available at: www.cnbc.com/2017/08/24/technology-killing-off-corporations-average-lifespan-of-company-under-20-years.html

G. West, *Scale: The Universal Laws of Growth, Innovation, Sustainability, and the Pace of Life in Organisms, Cities, Economies, and Companies*, Penguin Press, 2017.

Chapter 14

M. Roser, E. Ortiz-Ospina, and H. Ritchie, "Life expectancy", *Our World in Data*, 2013. Available at: ourworldindata.org/life-expectancy

M. Roser, H. Ritchie, and E. Ortiz-Ospina, "World population growth", *Our World in Data*, 2013. Available at: ourworldindata.org/world-population-growth

H. Rosling, A. R. Rönnlund, and O. Rosling, *Factfulness: Ten Reasons We're Wrong About the World – and Why Things Are Better Than You Think*, Flatiron Books, 2018.

V. Smil, *Growth*, The MIT Press, 2019.

S. H. Woolf and H. Schoomaker, "Life expectancy and mortality rates in the United States, 1959–2017", *JAMA*, 322 (20) 2019.

World Bank, "Fertility rate, total (births per woman) – Japan, Korea, Rep". Available at: data.worldbank.org/indicator/SP.DYN.TFRT.IN?locations=JP-KR

World Health Organization, "Life expectancy", *Global Health Observatory data*. Available at: www.who.int/gho/mortality_burden_disease/life_tables/situation_trends_text/en/

Chapter 15

R. Angel, "Feasibility of cooling the Earth with a cloud of small spacecraft near the inner Lagrange point (L1)", *Proceedings of the National Academy of Sciences*, 103 (46), 2006.

D. Dunne, "Explainer: six ideas to limit global warming with solar geoengineering", *Carbon Brief*, 2018. Available at: www.carbonbrief.org/explainer-six-ideas-to-limit-global-warming-with-solar-geoengineering

A. Gabbatt and agencies, "IBM computer Watson wins Jeopardy clash", *The Guardian*, 2011.

A. Grubler et al, "A low energy demand scenario for meeting the 1.5°C target and sustainable development goals without negative emission technologies", *Nature Energy*, 3 (6), 2018.

International Energy Agency, "Offshore wind outlook 2019". Available at: www.iea.org/reports/offshore-wind-outlook-2019

Oxford Economics, "How robots change the world", 2019. Available at: resources.oxfordeconomics.com/how-robots-change-the-world

D. Silver et al, "A general reinforcement learning algorithm that masters chess, shogi, and Go through self-play", *Science*, 362 (6419), 2018.

B. J. Soden et al, "Global cooling after the eruption of Mount Pinatubo: a test of climate feedback by water vapor", *Science*, 296 (5568), 2002.

M. Tegmark, *Life 3.0: Being Human in the Age of Artificial Intelligence*, Deckle Edge, 2017.

"What is 5G and what will it mean for you?", *BBC News*, 2020.

Chapter 16

K. W. Bandilla, "Carbon capture and storage", *Future Energy*, Elsevier, 2020.

M. De Wit et al, *The Circularity Gap Report 2020*, Circle Economy, 2020.

J. Mercure et al, "Macroeconomic impact of stranded fossil fuel assets", *Nature Climate Change*, 8 (7), 2018.

J. Pretty et al, "Global assessment of agricultural system redesign for sustainable intensification", *Nature Sustainability*, 1 (8), 2018.

Chapter 17

IRENA, "Measuring the socio-economics of transition: Focus on jobs", International Renewable Energy Agency, 2020. Available at: www.irena.org/publications/2020/Feb/Measuring-the-socioeconomics-of-transition-Focus-on-jobs

Chapter 18

P. Ball, "The new history", *Nature*, 480 (7378), 2011.

D. Centola et al, "Experimental evidence for tipping points in social convention", *Science*, 360 (6393), 2018.

B. Ewers et al, "Divestment may burst the carbon bubble if investors' beliefs tip to anticipating strong future climate policy", arXiv:1902.07481, 2019.

J. Falk and O. Gaffney et al, "Exponential climate action roadmap", Future Earth, 2018. Available at: exponentialroadmap.org/wp-content/uploads/2018/09/Exponential-Climate-Action-Roadmap-September-2018.pdf

D. F. Lawson et al, "Children can foster climate change concern among their parents", *Nature Climate Change*, 9 (6), 2019.

A. Leiserowitz et al, *Climate Change in the American Mind: November 2019*, Yale Program on Climate Change Communication, 2019. Available at: climatecommunication.yale.edu/wp-content/uploads/2019/12/Climate_Change_American_Mind_November_2019b.pdf

PricewaterhouseCoopers, "Navigating the rising tide of uncertainty", 23, 2020. Available at: www.pwc.com/gx/en/ceo-agenda/ceosurvey/2020.html

M. Taylor, J. Watts, and J. Bartlett, "Climate crisis: 6 million people join latest wave of global protests", *The Guardian*, 2019.

J. Watts, "Greta Thunberg, schoolgirl climate change warrior: 'Some people can let things go. I can't'", *The Guardian*, 2019.

YouGov, "Climate change protesters have been disrupting roads and public transport, aiming to 'shut down London' in order to bring attention the their cause. Do you support or oppose these actions?", 2010. Available at: yougov.co.uk/topics/science/survey-results/daily/2019/04/17/35ede/1

"1000+ Divestment Commitments", *Fossil Free: Divestment*. Available at: gofossilfree.org/divestment/commitments/

Chapter 19

D. Grinspoon, *Earth in Human Hands: Shaping our Planet's Future*, Grand Central Publishing, 2016.

INDEX

ACKNOWLEDGMENTS

This book would not have been possible without the help, encouragement, and wisdom of many, many people. Special thanks go to Félix Pharand-Deschênes, who has done more than anyone else to visualize the Anthropocene; Becky Gee, our editor at DK, and Kaela Slavik, our research assistant; John Ash at Pew; Peter Kindersley and the team at DK: Angeles Gavira, Michael Duffy, Jonathan Metcalf, and Liz Wheeler; the Silverback Films team: Jon Clay, Colin Butfield, Claire Sharrock, Alistair Fothergill, Keith Scholey, and Ana Taboada; Will Steffen, Matteo Willeit, and Denise Young, in particular, and colleagues at the Potsdam Institute for Climate Impact Research and the Stockholm Resilience Centre, in general, for scientific guidance; the research team of my (Johan) European Research Council project – Earth Resilience in the Anthropocene (ERA) – coordinated by Sarah Cornell and Jonathan Donges, for providing latest understanding on Earth resilience; and the planetary emergency core team: Sandrine Dixson-Declève, James Lloyd, Bernadette Fischler, and Elise Buckle. Special thanks also to Mark Prain of the Edmund Hillary Fellowship in New Zealand, who inspired us to write the "best untold story in town", which eventually became this book; and to our families: George, Oscar, and Sophie on Owen's side and Vera, Alex, Isak, and Ulrika on Johan's side.

PICTURE CREDITS

A1 Globaïa: data generated using auto-RIFT and provided by the NASA MEaSUREs ITS_LIVE project; A. M. Le Brocq et al, "Evidence from ice shelves for channelized meltwater flow beneath the Antarctic Ice Sheet", *Nature Geoscience*, 6(11), 2013
A2/3 Globaïa: adapted from Burke et al, *PNAS*, 2018. See ww.pnas.org/content/115/52/13288
A4 Globaïa: adapted from the Earth Commission of the Global Commons Alliance
B1 Globaïa: data sourced from Hansen/UMD/Google/USGS/NASA; M. C. Hansen et al, "High-resolution global maps of 21st-century forest cover change", *Science*, 342, 2013
B2/3 Globaïa: adapted from C. M. Kennedy et al, "Managing the middle: A shift in conservation priorities based on the global human modification gradient", *Global Change Biology*, 25 (3), 2019. See doi.org/10.1111/gcb.14549
B4 Globaïa: adapted from J. Rockström et al, "A safe operating space for humanity", *Nature*, 461 (7263), 2009; W. Steffen et al, "Planetary boundaries: Guiding human development on a changing planet", *Science*, 347 (6223), 2015
C1 Globaïa: for a full list of data included in this visualization, see www.globaia.org
C2/3 Globaïa
C4 Globaïa: adapted from W. Steffen et al, "The trajectory of the Anthropocene: The Great Acceleration", *Anthropocene Review*, 2015
D1 Globaïa: upper panel, adapted from Stockholm Resilience Centre/Azote graphic; lower panel, adapted from The World in 2050 Report:

Transformations to achieve the Sustainable Development Goals, published by the International Institute for Applied Systems, 2018
D2 Globaïa: adapted from H. Rosling et al, *Factfulness*, Flatiron Books, 2018
D3 Globaïa: adapted from E. Dinerstein et al, *Science Advances*, 2020
D4 Globaïa
p25 © Dorling Kindersley
p37 Adapted from C. MacFarling Meure et al, 2006, and D. Lüthi et al, 2008, for the data underlying the figure. "High-resolution carbon dioxide concentration record 650,000–800,000 years before present", *Nature*, 453, 379–382, 15 May 2008
p45 S. Montgomery, "Hominin brain evolution: The only way is up?, *Current Biology*, 2016. See doi.org/10.1016/j.cub.2018.06.021
p55 Adapted from Simon L. Lewis and Mark A. Maslin's *The Human Planet: How We Created the Anthropocene*, Penguin, 2018
p67 T. Juniper, *What's Really Happening to our Planet?*, Dorling Kindersley, 2016
p93 Adapted from Will Steffen, Johan Rockström et al, "Trajectories of the Earth system in the Anthropocene", *Proceedings of the National Academy of Sciences*, 115 (33), 8252–8259, Aug 2018; DOI: 10.1073/pnas.1810141115
p100 Adapted from T. M. Lenton et al, "Climate tipping points: Too risky to bet against", *Nature*, 575, 592–595, 2019. See www.nature.com/articles/d41586-019-03595-0
p110 Adapted from D. Meadows, "Leverage points: Places to intervene in a system", 1999. Hartland, WI: The Sustainability Institute

p125 Adapted from J. Rockström, O. Gaffney et al, "A roadmap for rapid decarbonisation", *Science*, 2017
p135 Adapted from J. Poore and T. Nemecek, "Reducing food's environmental impacts through producers and consumers", *Science*, 360 (6392), 987–992, with additional calculations by Our World in Data. See ourworldindata.org/grapher/land-use-protein-poore
p149 Adapted from Richard G. R. Wilkinson and Kate Pickett, *The Spirit Level: Why More Equal Societies Almost Always Do Better*, Allen Lane, 2009
p158 Adapted from UN World Urbanization Prospects, 2018. See ourworldindata.org/grapher/urban-and-rural-population
p167 Our World in Data, based on HYDE, UN and UN Population Division [2019 Revision]. See ourworldindata.org/future-population-growth
p174 Adapted from The Natural Edge Project, 2004, Griffith University, and Australian National University, Australia
p185 Adapted from K. Raworth, "A Doughnut for the Anthropocene: Humanity's compass in the 21st century", *The Lancet: Planetary Health*, 1, 2017
p203 Adapted from World Values Survey (2014), sourced from Our World in Data. See ourworldindata.org/grapher/self-reported-trust-attitudes?country=CHN~FIN~NZL~NOR~SWE
p206 © Dorling Kindersley